FTCE Mathematics

Teacher Certification Exam

By: Sharon Wynne, M.S
Southern Connecticut State University

"And, while there's no reason yet to panic, I think it's only prudent that we make preparations to panic."

XAMonline, INC.
Boston

Copyright © 2008 XAMonline, Inc.
All rights reserved. No part of the material protected by this copyright notice may be reproduced or utilized in any form or by any means, electronic or mechanical, including photocopying, recording or by any information storage and retrievable system, without written permission from the copyright holder.

To obtain permission(s) to use the material from this work for any purpose including workshops or seminars, please submit a written request to:

> XAMonline, Inc.
> 21 Orient Ave.
> Melrose, MA 02176
> Toll Free 1-800-509-4128
> Email: info@xamonline.com
> Web www.xamonline.com
> Fax: 1-781-662-9268

Library of Congress Cataloging-in-Publication Data

Wynne, Sharon A.
    Mathematics 6-12: Teacher Certification / Sharon A. Wynne. -2nd ed.
    ISBN 978-1-58197-640-3
    1. Mathematics 6-12.   2. Study Guides.   3. FTCE
    4. Teachers' Certification & Licensure.   5. Careers

Disclaimer:

The opinions expressed in this publication are the sole works of XAMonline and were created independently from the National Education Association, Educational Testing Service, or any State Department of Education, National Evaluation Systems or other testing affiliates.

Between the time of publication and printing, state specific standards as well as testing formats and website information may change that is not included in part or in whole within this product. Sample test questions are developed by XAMonline and reflect similar content as on real tests; however, they are not former tests. XAMonline assembles content that aligns with state standards but makes no claims nor guarantees teacher candidates a passing score. Numerical scores are determined by testing companies such as NES or ETS and then are compared with individual state standards. A passing score varies from state to state.

Printed in the United States of America

FTCE: Mathematics 6-12
ISBN: 978-1-58197-640-3

TEACHER CERTIFICATION EXAM

About the Subject Assessments

FTCE™: Subject Assessment in the Mathematics 6-12 examination

Purpose: The assessments are designed to test the knowledge and competencies of prospective secondary level teachers. The question bank from which the assessment is drawn is undergoing constant revision. As a result, your test may include questions that will not count towards your score.

Test Version: There are two versions of subject assessment for Mathematics examination in Florida. The Middle Grades Mathematics 5-9 exam emphasizes comprehension in Knowledge of mathematics through problem solving; Knowledge of mathematical representations; Knowledge of mathematics through reasoning; Knowledge of mathematical connections; Knowledge of number sense, concepts, and operations; Knowledge of algebraic thinking; Knowledge of data analysis and probability; Knowledge of geometry and spatial sense; Knowledge of measurement. The Mathematics 6-12 exam emphasizes comprehension in Knowledge of Algebra; Knowledge of Functions; Knowledge of Geometry from a Synthetic Perspective; Knowledge of Geometry from an Algebraic Perspective; Knowledge of Trigonometry; Knowledge of Statistics; Knowledge of Probability; Knowledge of Discrete Mathematics; Knowledge of Calculus; Knowledge of Number Sense and Mathematical Structure; Knowledge of Mathematics as Communication; Knowledge of Mathematics as Reasoning; Knowledge of Mathematical Connections; Knowledge of Instruction; Knowledge of Assessment. The Mathematics 6-12 study guide is based on a typical knowledge level of persons who have completed a <u>*bachelor's degree program*</u> in Mathematics.

Time Allowance: You will have 2½ hours to finish the exam.

Additional Information about the FTCE Assessments: The FTCE™ series subject assessments are developed *Florida Department of Education* of Tallahassee, FL. They provide additional information on the FTCE series assessments, including registration, preparation and testing procedures and study materials such topical guides.

TEACHER CERTIFICATION EXAM

TABLE OF CONTENTS

COMPETENCY/SKILL   **PG #**

1.0 KNOWLEDGE OF ALGEBRA 1

    1.1. Identify graphs of linear inequalities on a number line 1

    1.2. Identify graphs of linear equations and inequalities in the coordinate plane 2

    1.3. Identify or interpret the slope and intercepts of a linear graph or a linear equation 4

    1.4. Determine the equation of a line, given the appropriate information such as two points, point-slope, slope-intercept, or its graph 5

    1.5. Solve problems involving the use of equations containing rational algebraic expressions 7

    1.6. Factor polynomials (e.g., the sum or difference of two cubes) 13

    1.7. Rewrite radical and rational expressions into equivalent forms 14

    1.8. Perform the four basic operations on rational and radical expressions 18

    1.9. Solve equations containing radicals 21

    1.10. Multiply or divide binomials containing radicals 23

    1.11. Solve quadratic equations by factoring, graphing, completing the square, or using the quadratic formula, including complex solutions 24

    1.12. Solve problems using quadratic equations 26

    1.13. Use the discriminant to determine the nature of solutions of quadratic equations 29

    1.14. Determine a quadratic equation from known roots 30

    1.15. Identify the graphs of quadratic inequalities 32

    1.16. Solve real-world problems using direct and inverse variations 33

    1.17. Solve systems of linear equations or inequalities 34

    1.18. Solve systems of linear inequalities graphically 45

MATHEMATICS 6-12

1.19. Formulate or identify systems of linear equations or inequalities to solve real-world problems 45

1.20. Solve equations or inequalities involving absolute value 49

1.21. Expand given binomials to a specified positive integral power 51

1.22. Determine a specified term in the expansion of given binomials 52

1.23. Solve polynomial equations by factoring 53

1.24. Perform vector addition, subtraction, and scalar multiplication on the plane 55

1.25. Solve real-world problems involving ratio or proportion 57

2.0 KNOWLEDGE OF FUNCTIONS 59

2.1. Interpret the language and notation of functions 59

2.2. Determine which relations are functions, given mappings, sets of ordered pairs, rules, and graphs 59

2.3. Identify the domain and range of a given function 62

2.4. Identify the graph of special functions (i.e., absolute value, step, piecewise, identity, constant function) 64

2.5. Find specific values of a given function 69

2.6. Estimate or find the zeros of a polynomial function 74

2.7. Identify the sum, difference, product, and quotient of functions 77

2.8. Determine the inverse of a given function 78

2.9. Determine the composition of two functions 79

2.10. Determine whether a function is symmetric, periodic, or even/odd 80

2.11. Determine the graph of the image of a function under given transformations (i.e., translation, rotations through multiples of 90 degrees, dilations, and/or reflections over y=x horizontal or vertical lines) 81

MATHEMATICS 6-12

TEACHER CERTIFICATION EXAM

3.0 KNOWLEDGE OF GEOMETRY FROM A SYNTHETIC PERSPECTIVE 83

    3.1. Determine the change in the area or volume of a figure when its dimensions are altered 83

    3.2. Estimate measurements of familiar objects using metric or standard units 84

    3.3. Determine the relationships between points, lines, and planes, including their intersections 85

    3.4. Classify geometric figures (e.g., lines, planes, angles, polygons, solids) according to their properties 86

    3.5. Determine the measures of interior and exterior angles of any polygon 89

    3.6. Determine the sum of the measures of the interior angles and the sum of the measures of the exterior angles of convex polygons 90

    3.7. Identify applications of special properties of trapezoids, parallelograms, and kites 91

    3.8. Solve problems using the definition of congruent polygons and related theorems 93

    3.9. Solve problems using the definition of similar polygons and solids and related theorems 98

    3.10. Apply the Pythagorean theorem or its converse 102

    3.11. Use 30-60-90 or 45-45-90 triangle relationships to determine the lengths of the sides of triangles 103

    3.12. Calculate the perimeter, circumference, and/or area of two-dimensional figures (e.g., circles, sectors, polygons, composite figures) 104

    3.13. Apply the theorems pertaining to the relationships of chords, secants, diameters, radii, and tangents with respect to circles and to each other .. 110

    3.14. Apply the theorems pertaining to the measures of inscribed angles and angles formed by chords, secants, and tangents 113

    3.15. Identify basic geometric constructions (e.g., bisecting angles or line segments, constructing parallels or perpendiculars) 115

TEACHER CERTIFICATION EXAM

3.16. Identify the converse, inverse, and contrapositive of a conditional statement 123

3.17. Identify valid conclusions from given statements 124

3.18. Classify examples of reasoning processes as inductive or deductive 127

3.19. Determine the surface area and volume of prisms, pyramids, cylinders, cones, and spheres 128

3.20. Identify solids and their related nets 132

4.0 KNOWLEDGE OF GEOMETRY FROM AN ALGEBRAIC PERSPECTIVE 133

4.1. Solve distance and midpoint problems involving two points, a point and a line, two lines, and two parallel lines 133

4.2. Identify the directrix, foci, vertices, axes, and asymptotes of a conic section where appropriate 137

4.3. Determine the center and the radius of a circle given its equation, and identify the graph 142

4.4. Identify the equation of a conic section, given the appropriate information 144

4.5. Use translations, rotations, dilations, or reflections on a coordinate plane to identify the images of geometric objects under such transformations 147

5.0 KNOWLEDGE OF TRIGONOMETRY 149

5.1. Identify equations of graphs of circular/trigonometric functions and their inverses 149

5.2. Solve problems involving circular/trigonometric function identities 150

5.3. Interpret the graphs of trigonometric functions (e.g., amplitude, period, phase shift) 153

5.4. Solve real-world problems involving triangles using the law of sines or the law of cosines 155

5.5. Use tangent, sine, and cosine ratios to solve right triangle problems 158

MATHEMATICS 6-12

TEACHER CERTIFICATION EXAM

6.0 KNOWLEDGE OF STATISTICS 160

    6.1. Interpret graphical data involving measures of location (i.e., percentiles, stanines, quartiles) 160

    6.2. Compute the mean, median, and mode of a set of data 161

    6.3. Determine whether the mean, the median, or the mode is the most appropriate measure of central tendency in a given situation 162

    6.4. Interpret the ranges, variances, and standard deviations for ungrouped data 163

    6.5. Interpret information from bar, line, picto-, and circle graphs; stem-and-leaf and scatter plots; and box-and-whisker graphs 164

    6.6. Interpret problems involving basic statistical concepts such as sampling, experimental design, correlation, and linear regression 170

7.0 KNOWLEDGE OF PROBABILITY 173

    7.1. Determine probabilities of dependent or independent events 173

    7.2. Predict odds of a given outcome 173

    7.3. Identify an appropriate sample space for an experiment 174

    7.4. Make predictions that are based on relative frequency of an event 175

    7.5. Determine probabilities using counting procedures, tables, tree diagrams, and formulas for permutations and combinations 176

8.0 KNOWLEDGE OF DISCRETE MATHEMATICS 182

    8.1. Find a specified term in an arithmetic sequence 182

    8.2. Find a specified term in a geometric sequence. 183

    8.3. Determine the sum of terms in an arithmetic or geometric progression 183

    8.4. Solve problems involving permutations and combinations 185

    8.5. Evaluate matrix expressions involving sums, differences, and products 186

    8.6. Rewrite a matrix equation as an equivalent system of linear equations or vice versa 191

TEACHER CERTIFICATION EXAM

8.7. Represent problem situations using discrete structures such as sequences, finite graphs, and matrices 192

9.0 KNOWLEDGE OF CALCULUS 195

9.1. Solve problems using the limit theorems concerning sums, products, and quotients of functions 195

9.2. Find the derivatives of algebraic, trigonometric, exponential, and logarithmic functions 196

9.3. Find the derivative of the sum, product, quotient, or the composition of functions 201

9.4. Identify and apply definitions of the derivative of a function 204

9.5. Use the derivative to find the slope of a curve at a point 205

9.6. Find the equation of a tangent line or a normal line at a point on a curve .. 206

9.7. Determine if a function is increasing or decreasing by using the first derivative in a given interval 209

9.8. Find relative and absolute maxima and minima 210

9.9. Find intervals on a curve where the curve is concave up or concave down 211

9.10. Identify points of inflection 212

9.11. Solve problems using velocity and acceleration of a particle moving along a line 214

9.12. Solve problems using instantaneous rates of change and related rates of change, such as growth and decay 216

9.13. Find antiderivatives for algebraic, trigonometric, exponential, and logarithmic functions 217

9.14. Solve distance, area, and volume problems using integration 222

9.15. Evaluate an integral by use of the fundamental theorem of calculus 228

MATHEMATICS 6-12

10.0 KNOWLEDGE OF NUMBER SENSE AND MATHEMATICAL STRUCTURE 229

10.1. Apply the properties of real numbers: closure, commutative, associative, distributive, identities, and inverses 229

10.2. Distinguish relationships between the complex number system and its subsystems......230

10.3. Apply inverse operations to solve problems (e.g., roots vs. powers, exponents vs. logarithms)......232

10.4. Apply number theory concepts (e.g., primes, factors, multiples) in real-world and mathematical problem situations......232

10.5. Identify numbers written in scientific notation, including the format used on scientific calculators and computers 237

11.0 KNOWLEDGE OF MATHEMATICS AS COMMUNICATION......239

11.1. Identify statements that correctly communicate mathematical definitions or concepts 239

11.2. Interpret written presentations of mathematics 239

11.3. Select or interpret appropriate concrete examples, pictorial illustrations, and symbolic representations in developing mathematical concepts 240

12.0 KNOWLEDGE OF MATHEMATICS AS REASONING 242

12.1. Identify reasonable conjectures 242

12.2. Identify a counter example to a conjecture 242

12.3. Identify simple valid arguments according to the laws of logic......243

12.4. Identify proofs for mathematical assertions, including direct and indirect proofs, proofs by mathematical induction, and proofs on a coordinate plane......245

12.5. Identify process skills: induction, deduction, questioning techniques, and observation-inference 247

TEACHER CERTIFICATION EXAM

13.0 KNOWLEDGE OF MATHEMATICAL CONNECTIONS 249

    13.1. Identify equivalent representations of the same concept or procedure (e.g., graphical, algebraic, verbal, numeric) 249

    13.2. Interpret relationships between mathematical topics (e.g., multiplication as repeated addition, powers as repeated multiplication) 249

    13.3. Interpret descriptions, diagrams, and representations of arithmetic operations 250

14.0 KNOWLEDGE OF INSTRUCTION 252

    14.1. Select appropriate resources for a classroom activity (e.g., manipulatives, mathematics models, technology, other teaching tools) ... 252

    14.2. Identify methods and strategies for teaching problem-solving skills and applications (e.g., constructing tables from given data, guess-and-check, working backwards, reasonableness, estimation) 252

15.0 KNOWLEDGE OF ASSESSMENT 255

    15.1. Identify students' errors, including multiple errors that result in correct or incorrect answers (e.g., algorithms, properties, drawings, procedures) 255

    15.2. Identify appropriate alternative methods of assessment (e.g., performance, portfolios, projects) 256

CURRICULUM AND INSTRUCTION 257

ANSWER KEY TO PRACTICE PROBLEMS 267

SAMPLE TEST 275

ANSWER KEY 289

RIGOR TABLE 290

RATIONALE 291

TEACHER CERTIFICATION STUDY GUIDE

Great Study and Testing Tips!

What to study in order to prepare for the subject assessments is the focus of this study guide but equally important is *how* you study.

You can increase your chances of truly mastering the information by taking some simple, but effective steps.

Study Tips:

1.   Some foods aid the learning process.   Foods such as milk, nuts, seeds, rice, and oats help your study efforts by releasing natural memory enhancers called CCKs (*cholecystokinin*) composed of *tryptophan*, *choline*, and *phenylalanine*. All of these chemicals enhance the neurotransmitters associated with memory. Before studying, try a light, protein-rich meal of eggs, turkey, and fish. All of these foods release the memory enhancing chemicals. The better the connections, the more you comprehend.

Likewise, before you take a test, stick to a light snack of energy boosting and relaxing foods. A glass of milk, a piece of fruit, or some peanuts all release various memory-boosting chemicals and help you to relax and focus on the subject at hand.

2.   Learn to take great notes.   A by-product of our modern culture is that we have grown accustomed to getting our information in short doses (i.e. TV news sound bites or USA Today style newspaper articles.)

Consequently, we've subconsciously trained ourselves to assimilate information better in neat little packages. If your notes are scrawled all over the paper, it fragments the flow of the information. Strive for clarity. Newspapers use a standard format to achieve clarity. Your notes can be much clearer through use of proper formatting. A very effective format is called *"Cornell Method."*

Take a sheet of loose-leaf lined notebook paper and draw a line all the way down the paper about 1-2" from the left-hand edge. Draw another line across the width of the paper about 1-2" up from the bottom. Repeat this process on the reverse side of the page.

Look at the highly effective result. You have ample room for notes, a left hand margin for special emphasis items or inserting supplementary data from the textbook, a large area at the bottom for a brief summary, and a little rectangular space for just about anything you want.

MATHEMATICS 6-12

3. **Get the concept then the details.** Too often we focus on the details and don't gather an understanding of the concept. However, if you simply memorize only dates, places, or names, you may well miss the whole point of the subject.

A key way to understand things is to put them in your own words. If you are working from a textbook, automatically summarize each paragraph in your mind. If you are outlining text, don't simply copy the author's words.

Rephrase them in your own words. You remember your own thoughts and words much better than someone else's, and subconsciously tend to associate the important details to the core concepts.

4. **Ask Why?** Pull apart written material paragraph by paragraph and don't forget the captions under the illustrations.

Example: If the heading is "Stream Erosion", flip it around to read "Why do streams erode?" Then answer the questions.

If you train your mind to think in a series of questions and answers, not only will you learn more, but it also helps to lessen the test anxiety because you are used to answering questions.

5. **Read for reinforcement and future needs.** Even if you only have 10 minutes, put your notes or a book in your hand. Your mind is similar to a computer; you have to input data in order to have it processed. *By reading, you are creating the neural connections for future retrieval.* The more times you read something, the more you reinforce the learning of ideas.

Even if you don't fully understand something on the first pass, *your mind stores much of the material for later recall.*

6. **Relax to learn, so go into exile.** Our bodies respond to an inner clock called biorhythms. Burning the midnight oil works well for some people, but not everyone.

If possible, set aside a particular place to study that is free of distractions. Shut off the television, cell phone, pager and exile your friends and family during your study period.

If you really are bothered by silence, try background music. Light classical music at a low volume has been shown to aid in concentration over other types.

Music that evokes pleasant emotions without lyrics are highly suggested. Try just about anything by Mozart. It relaxes you.

7.   Use arrows not highlighters. At best, it's difficult to read a page full of yellow, pink, blue, and green streaks.

Try staring at a neon sign for a while and you'll soon see my point, the horde of colors obscure the message.

A quick note, a brief dash of color, an underline, and an arrow pointing to a particular passage is much clearer than a horde of highlighted words.

8.   Budget your study time. Although you shouldn't ignore any of the material, *allocate your available study time in the same ratio that topics may appear on the test.*

TEACHER CERTIFICATION STUDY GUIDE

Testing Tips:

1. **Get smart, play dumb.** Don't read anything into the question. Don't make an assumption that the test writer is looking for something else than what is asked. Stick to the question as written and don't read extra things into it.

2. **Read the question and all the choices *twice* before answering the question.** You may miss something by not carefully reading, and then rereading both the question and the answers.

If you really don't have a clue as to the right answer, leave it blank on the first time through. Go on to the other questions, as they may provide a clue as to how to answer the skipped questions.

If later on, you still can't answer the skipped ones . . . *Guess.* The only penalty for guessing is that you *might* get it wrong. Only one thing is certain; if you don't put anything down, you will get it wrong!

3. **Turn the question into a statement.** Look at the way the questions are worded. The syntax of the question usually provides a clue. Does it seem more familiar as a statement rather than as a question? Does it sound strange?

By turning a question into a statement, you may be able to spot if an answer sounds right, and it may also trigger memories of material you have read.

4. **Look for hidden clues.** It's actually very difficult to compose multiple-foil (choice) questions without giving away part of the answer in the options presented.

In most multiple-choice questions you can often readily eliminate one or two of the potential answers. This leaves you with only two real possibilities and automatically your odds go to Fifty-Fifty for very little work.

5. **Trust your instincts.** For every fact that you have read, you subconsciously retain something of that knowledge. On questions that you aren't really certain about, go with your basic instincts, **your first impression on how to answer a question is usually correct.**

6. **Mark your answers directly on the test booklet.** Don't bother trying to fill in the optical scan sheet on the first pass through the test. *Just be very careful not to miss-mark your answers when you eventually transcribe them to the scan sheet.*

7. **Watch the clock!** You have a set amount of time to answer the questions. Don't get bogged down trying to answer a single question at the expense of 10 questions you can more readily answer.

MATHEMATICS 6-12

COMPETENCY 1.0 KNOWLEDGE OF ALGEBRA

SKILL 1.1 Identify graphs of linear inequalities on a number line.

- When graphing a first-degree equation, solve for the variable. The graph of this solution will be a single point on the number line. There will be no arrows.

- When graphing a linear inequality, the dot will be hollow if the inequality sign is $<$ or $>$. If the inequality signs is either $\geq$ or $\leq$, the dot on the graph will be solid. The arrow goes to the right for $\geq$ or $>$. The arrow goes to the left for $\leq$ or $<$.

Solve:

$$5(x+2)+2x = 3(x-2)$$
$$5x+10+2x = 3x-6$$
$$7x+10 = 3x-6$$
$$4x = {}^-16$$
$$x = {}^-4$$

Solve:

$$2(3x-7) > 10x-2$$
$$6x-14 > 10x-2$$
$${}^-4x > 12$$
$$x < {}^-3 \quad \text{Note the change in inequality when dividing by negative numbers.}$$

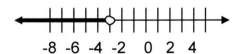

Solve the following equations and inequalities. Graph the solution set.

1. $5x - 1 > 14$
2. $7(2x - 3) + 5x = 19 - x$
3. $3x + 42 \geq 12x - 12$
4. $5 - 4(x + 3) = 9$

SKILL 1.2 Identify graphs of linear equations and inequalities in the coordinate plane.

A first degree equation has an equation of the form $ax + by = c$. To graph this equation, find either one point and the slope of the line or find two points. To find a point and slope, solve the equation for y. This gets the equation in **slope intercept form**, $y = mx + b$. The point (0,b) is the y-intercept and m is the line's slope. To find any 2 points, substitute any 2 numbers for x and solve for y. To find the intercepts, substitute 0 for x and then 0 for y.

Remember that graphs will go up as they go to the right when the slope is positive. Negative slopes make the lines go down as they go to the right.

If the equation solves to **x = any number**, then the graph is a **vertical line**. It only has an x intercept. Its slope is **undefined**.

If the equation solves to **y = any number**, then the graph is a **horizontal line**. It only has a y intercept. Its slope is 0 (zero).

When graphing a linear inequality, the line will be dotted if the inequality sign is $<$ or $>$. If the inequality signs are either $\geq$ or $\leq$, the line on the graph will be a solid line. Shade above the line when the inequality sign is $\geq$ or $>$. Shade below the line when the inequality sign is $<$ or $\leq$. Inequalities of the form $x >, x \leq, x <,$ or $x \geq$ number, draw a vertical line (solid or dotted). Shade to the right for $>$ or $\geq$. Shade to the left for $<$ or $\leq$. Remember: **Dividing or multiplying by a negative number will reverse the direction of the inequality sign.**

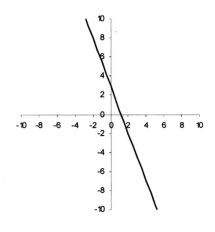

$5x + 2y = 6$

$y = {}^-5/2\, x + 3$

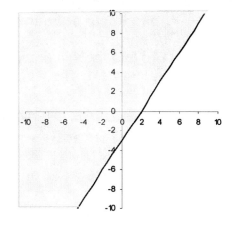

$3x - 2y \geq 6$

$y \leq 3/2\, x - 3$

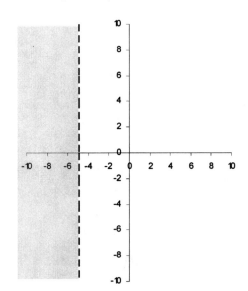

$3x + 12 < -3$

$x < {}^-5$

Graph the following:

1. $2x - y = {}^-4$
2. $x + 3y > 6$
3. $3x + 2y \leq 2y - 6$

SKILL 1.3 **Identify or interpret the slope and intercepts of a linear graph or a linear equation.**

To find the y intercept, substitute 0 for x and solve for y. This is the y intercept. The y intercept is also the value of b in $y = mx + b$.

To find the x intercept, substitute 0 for y and solve for x. This is the x intercept.

1. Find the slope and intercepts of $3x + 2y = 14$.

$$3x + 2y = 14$$
$$2y = {}^-3x + 14$$
$$y = {}^-3/2\ x + 7$$

The slope of the line is $^-3/2$, the value of m.
The y intercept of the line is 7.

The intercepts can also be found by substituting 0 in place of the other variable in the equation.

To find the y intercept:	To find the x intercept:
let $x = 0$; $3(0) + 2y = 14$	let $y = 0$; $3x + 2(0) = 14$
$0 + 2y = 14$	$3x + 0 = 14$
$2y = 14$	$3x = 14$
$y = 7$	$x = 14/3$
$(0,7)$ is the y intercept.	$(14/3, 0)$ is the x intercept.

Find the slope and the intercepts (if they exist) for these equations:

1. $5x + 7y = {}^-70$
2. $x - 2y = 14$
3. $5x + 3y = 3(5 + y)$
4. $2x + 5y = 15$

MATHEMATICS 6-12

SKILL 1.4 Determine the equation of a line, given the appropriate information such as two points, point-slope, slope-intercept, or its graph.

-The point-slope form for the equation of a line:

$$(y - y_1) = m(x - x_1)$$

where m is the slope and (x_1, y_1) is the point.

- The equation of a graph can be found by finding its slope and its y intercept. To find the slope, find 2 points on the graph where co-ordinates are integer values. Using points: (x_1, y_1) and (x_2, y_2).

$$\text{slope} = \frac{y_2 - y_1}{x_2 - x_1}$$

The y intercept is the y coordinate of the point where a line crosses the y axis. The equation can be written in slope-intercept form, which is $y = mx + b$, where m is the slope and b is the y intercept. To rewrite the equation into some other form, multiply each term by the common denominator of all the fractions. Then rearrange terms as necessary.

- Given two points on a line, the first thing to do is to find the slope of the line. If 2 points on the graph are (x_1, y_1) and (x_2, y_2), then the slope is found using the formula:

$$\text{slope} = \frac{y_2 - y_1}{x_2 - x_1}$$

The slope will now be denoted by the letter **m**. To write the equation of a line, choose either point. Substitute them into the formula:

$$Y - y_a = m(X - x_a)$$

Remember (x_a, y_a) can be (x_1, y_1) or (x_2, y_2) If **m**, the value of the slope, is distributed through the parentheses, the equation can be rewritten into other forms of the equation of a line.

Find the equation of a line through $(9, ^-6)$ and $(^-1, 2)$.

$$Cos63 = \frac{x}{12}$$

$Y - y_a = m(X - x_a) \to Y - 2 = {^-4/5}(X - {^-1}) \to$
$Y - 2 = {^-4/5}(X + 1) \to Y - 2 = {^-4/5}X - 4/5 \to$
$Y = {^-4/5}X + 6/5$ This is the slope-intercept form.

Multiplying by 5 to eliminate fractions, it is:
$5Y = {^-4}X + 6 \to 4X + 5Y = 6$ Standard form.

Write the equation of a line through these two points:
1. $(5, 8)$ and $(^-3, 2)$
2. $(11, 10)$ and $(11, ^-3)$
3. $(^-4, 6)$ and $(6, 12)$
4. $(7, 5)$ and $(^-3, 5)$

SKILL 1.5 Solve problems involving the use of equations containing rational algebraic expressions.

Add or subtract rational algebraic fractions.

- In order to add or subtract rational expressions, they must have a common denominator. If they don't have a common denominator, then factor the denominators to determine what factors are missing from each denominator to make the LCD. Multiply both numerator and denominator by the missing factor(s). Once the fractions have a common denominator, add or subtract their numerators, but keep the common denominator the same. Factor the numerator if possible and reduce if there are any factors that can be cancelled.

1. Find the least common denominator for $6a^3b^2$ and $4ab^3$.

These factor into $2 \cdot 3 \cdot a^3 \cdot b^2$ and $2 \cdot 2 \cdot a \cdot b^3$.
The first expression needs to be multiplied by another 2 and b.
The other expression needs to be multiplied by 3 and a^2.
Then both expressions would be
$2 \cdot 2 \cdot 3 \cdot a^3 \cdot b^3 = 12a^3b^3 = $ LCD.

2. Find the LCD for $x^2 - 4$, $x^2 + 5x + 6$, and $x^2 + x - 6$.

$x^2 - 4$ factors into $(x-2)(x+2)$
$x^2 + 5x + 6$ factors into $(x+3)(x+2)$
$x^2 + x - 6$ factors into $(x+3)(x-2)$

To make these lists of factors the same, they must all be $(x+3)(x+2)(x-2)$. This is the LCD.

3.

$$\frac{5}{6a^3b^2} + \frac{1}{4ab^3} = \frac{5(2b)}{6a^3b^2(2b)} + \frac{1(3a^2)}{4ab^3(3a^2)} = \frac{10b}{12a^3b^3} + \frac{3a^2}{12a^3b^3} = \frac{10b + 3a^2}{12a^3b^3}$$

This will not reduce as all 3 terms are not divisible by anything.

4.

$$\frac{2}{x^2-4} - \frac{3}{x^2+5x+6} + \frac{7}{x^2+x-6} =$$

$$\frac{2}{(x-2)(x+2)} - \frac{3}{(x+3)(x+2)} + \frac{7}{(x+3)(x-2)} =$$

$$\frac{2(x+3)}{(x-2)(x+2)(x+3)} - \frac{3(x-2)}{(x+3)(x+2)(x-2)} + \frac{7(x+2)}{(x+3)(x-2)(x+2)} =$$

$$\frac{2x+6}{(x-2)(x+2)(x+3)} - \frac{3x-6}{(x+3)(x+2)(x-2)} + \frac{7x+14}{(x+3)(x-2)(x+2)} =$$

$$\frac{2x+6-(3x-6)+7x+14}{(x+3)(x-2)(x+2)} = \frac{6x+26}{(x+3)(x-2)(x+2)}$$

This will not reduce.

Try These:

1. $\dfrac{6}{x-3} + \dfrac{2}{x+7}$

2. $\dfrac{5}{4a^2b^5} + \dfrac{3}{5a^4b^3}$

3. $\dfrac{x+3}{x^2-25} + \dfrac{x-6}{x^2-2x-15}$

Solve word problems with rational algebraic expressions and equations.

- Some problems can be solved using equations with rational expressions. First write the equation. To solve it, multiply each term by the LCD of all fractions. This will cancel out all of the denominators and give an equivalent algebraic equation that can be solved.

1. The denominator of a fraction is two less than three times the numerator. If 3 is added to both the numerator and denominator, the new fraction equals 1/2.

original fraction: $\dfrac{x}{3x-2}$ revised fraction: $\dfrac{x+3}{3x+1}$

$$\dfrac{x+3}{3x+1} = \dfrac{1}{2}$$

$$2x + 6 = 3x + 1$$

$$x = 5$$

original fraction: $\dfrac{5}{13}$

2. Elly Mae can feed the animals in 15 minutes. Jethro can feed them in 10 minutes. How long will it take them if they work together?

Solution: If Elly Mae can feed the animals in 15 minutes, then she could feed 1/15 of them in 1 minute, 2/15 of them in 2 minutes, $x/15$ of them in x minutes. In the same fashion Jethro could feed $x/10$ of them in x minutes. Together they complete 1 job. The equation is:

$$\dfrac{x}{15} + \dfrac{x}{10} = 1$$

Multiply each term by the LCD of 30:

$$2x + 3x = 30$$

$$x = 6 \text{ minutes}$$

MATHEMATICS 6-12

3. A salesman drove 480 miles from Pittsburgh to Hartford. The next day he returned the same distance to Pittsburgh in half an hour less time than his original trip took, because he increased his average speed by 4 mph. Find his original speed.

Since distance = rate x time then time = $\frac{\text{distance}}{\text{rate}}$

original time − 1/2 hour = shorter return time

$$\frac{480}{x} - \frac{1}{2} = \frac{480}{x+4}$$

Multiplying by the LCD of $2x(x+4)$, the equation becomes:

$$480[2(x+4)] - 1[x(x+4)] = 480(2x)$$
$$960x + 3840 - x^2 - 4x = 960x$$
$$x^2 + 4x - 3840 = 0$$
$$(x+64)(x-60) = 0 \quad \text{Either (x-60=0) or (x+64=0) or both=0}$$
$$x = 60 \qquad\qquad\qquad 60 \text{ mph is the original speed.}$$
$$x = 64 \qquad\qquad\qquad \text{This is the solution since the time}$$
cannot be negative. Check your answer

$$\frac{480}{60} - \frac{1}{2} = \frac{480}{64}$$
$$8 - \frac{1}{2} = 7\frac{1}{2}$$
$$7\frac{1}{2} = 7\frac{1}{2}$$

Try these:

1. Working together, Larry, Moe, and Curly can paint an elephant in 3 minutes. Working alone, it would take Larry 10 minutes or Moe 6 minutes to paint the elephant. How long would it take Curly to paint the elephant if he worked alone?

2. The denominator of a fraction is 5 more than twice the numerator. If the numerator is doubled, and the denominator is increased by 5, the new fraction is equal to 1/2. Find the original number.

3. A trip from Augusta, Maine to Galveston, Texas is 2108 miles. If one car drove 6 mph faster than a truck and got to Galveston 3 hours before the truck, find the speeds of the car and truck.

Solve rational algebraic equations for one variable.

- To solve an algebraic formula for some variable, called R, follow the following steps:

a. Eliminate any parentheses using the distributive property.
b. Multiply every term by the LCD of any fractions to write an equivalent equation without any fractions.
c. Move all terms containing the variable, R, to one side of the equation. Move all terms without the variable to the opposite side of the equation.
d. If there are 2 or more terms containing the variable R, factor **only R** out of each of those terms as a common factor.
e. Divide both sides of the equation by the number or expression being multiplied times the variable, R.
f. Reduce fractions if possible.
g. Remember there are restrictions on values allowed for variables because the denominator can not equal zero.

1. Solve $A = p + prt$ for t.

$$A - p = prt$$
$$\frac{A-p}{pr} = \frac{prt}{pr}$$
$$\frac{A-p}{pr} = t$$

2. Solve $A = p + prt$ for p.

$$A = p(1+rt)$$
$$\frac{A}{1+rt} = \frac{p(1+rt)}{1+rt}$$
$$\frac{A}{1+rt} = p$$

3. $A = 1/2 \; h(b_1 + b_2)$ for b_2

$$A = 1/2 \; hb_1 + 1/2 \; hb_2 \quad \leftarrow \text{step a}$$
$$2A = hb_1 + hb_2 \quad \leftarrow \text{step b}$$
$$2A - hb_1 = hb_2 \quad \leftarrow \text{step c}$$
$$\frac{2A - hb_1}{h} = \frac{hb_2}{h} \quad \leftarrow \text{step d}$$
$$\frac{2A - hb_1}{h} = b_2 \quad \leftarrow \text{will not reduce}$$

Solve:
1. $F = 9/5 \; C + 32$ for C
2. $A = 1/2 \; bh + h^2$ for b
3. $S = 180(n-2)$ for n

To solve an equation with rational expressions, find the least common denominator of all the fractions. Multiply each term by the LCD of all fractions. This will cancel out all of the denominators and give an equivalent algebraic equation that can be solved. Solve the resulting equation. Once you have found the answer(s), substitute them back into the original equation to check them. Sometimes there are solutions that do not check in the original equation. These are extraneous solutions, which are not correct and must be eliminated. If a problem has more than one potential solution, each solution must be checked separately.

NOTE: What this really means is that you can substitute the answers from any multiple choice test back into the question to determine which answer choice is correct.

Solve and **check**:

1. $\dfrac{72}{x+3} = \dfrac{32}{x+3} + 5 \qquad$ LCD $= x+3$, so multiply by this.

$(x+3) \times \dfrac{72}{x+3} = (x+3) \times \dfrac{32}{x+3} + 5(x+3)$

$72 = 32 + 5(x+3) \rightarrow 72 = 32 + 5x + 15$

$72 = 47 + 5x \qquad \rightarrow 25 = 5x$

$5 = x$ (This checks too).

2. $\dfrac{12}{2x^2-4x} + \dfrac{13}{5} = \dfrac{9}{x-2} \qquad$ Factor $2x^2 - 4x = 2x(x-2)$.

$\qquad\qquad\qquad\qquad\qquad\qquad$ LCD $= 5 \times 2x(x-2)$ or $10x(x-2)$

$10x(x-2) \times \dfrac{12}{2x(x-2)} + 10x(x-2) \times \dfrac{13}{5} = \dfrac{9}{x-2} \times 10x(x-2)$

$60 + 2x(x-2)(13) = 90x$

$26x^2 - 142x + 60 = 0$

$2(13x^2 - 71x + 30) = 0$

$2(x-5)(13x-6) \qquad$ so $x = 5$ or $x = 6/13 \;\; \leftarrow$ both check

Try these:

1. $\dfrac{x+5}{3x-5} + \dfrac{x-3}{2x+2} = 1$

2. $\dfrac{2x-7}{2x+5} = \dfrac{x-6}{x+8}$

SKILL 1.6 Factor polynomials (e.g., the sum or difference of two cubes).

Factor the sum or difference of two cubes

- To factor the sum or the difference of perfect cubes, follow this procedure:

a. Factor out any greatest common factor (GCF).

b. Make a parentheses for a binomial (2 terms) followed by a trinomial (3 terms).

c. The sign in the first parentheses is the same as the sign in the problem. The difference of cubes will have a "-" sign in the first parentheses. The sum of cubes will use a "+".

d. The first sign in the second parentheses is the opposite of the sign in the first parentheses. The second sign in the other parentheses is always a "+".

e. Determine what would be cubed to equal each term of the problem. Put those expressions in the first parentheses.

f. To make the 3 terms of the trinomial, think square - product - square. Looking at the binomial, square the first term. This is the trinomial's first term. Looking at the binomial, find the product of the two terms, ignoring the signs. This is the trinomial's second term. Looking at the binomial, square the third term. This is the trinomial's third term. Except in rare instances, the trinomial does not factor again.

Factor completely:

1.

$16x^3 + 54y^3$

$2(8x^3 + 27y^3)$ ← GCF

$2(\quad + \quad)(\quad - \quad + \quad)$ ← signs

$2(2x + 3y)(\quad - \quad + \quad)$ ← what is cubed to equal $8x^3$ or $27y^3$

$2(2x + 3y)(4x^2 - 6xy + 9y^2)$ ← square-product-square

2.

$64a^3 - 125b^3$

$(\quad - \quad)(\quad + \quad + \quad)$ ← signs

$(4a - 5b)(\quad + \quad + \quad)$ ← what is cubed to equal $64a^3$ or $125b^3$

$(4a - 5b)(16a^2 + 20ab + 25b^2)$ ← square-product-square

3.

$27x^{27} + 343y^{12} = (3x^9 + 7y^4)(9x^{18} - 21x^9y4 + 49y^8)$

Note: The coefficient 27 is different from the exponent 27.

Try These:

1. $216x^3 - 125y^3$
2. $4a^3 - 32b^3$
3. $40x^{29} + 135x^2y^3$

SKILL 1.7 Rewrite radical and rational expressions into equivalent forms.

Rational expressions can be changed into other equivalent fractions by either reducing them or by changing them to have a common denominator. When dividing any number of terms by a single term, divide or reduce their coefficients. Then subtract the exponent of a variable on the bottom from the exponent of the same variable from the numerator.

To reduce a rational expression with more than one term in the denominator, the expression must be factored first. Factors that are exactly the same will cancel and each become a 1. Factors that have exactly the opposite signs of each other, such as $(a - b)$ and $(b - a)$, will cancel and one factor becomes a 1 and the other becomes a $^-1$.

To make a fraction have a common denominator, factor the fraction. Determine what factors are missing from that particular denominator, and multiply both the numerator and the denominator by those missing factors. This gives a new fraction which now has the common denominator.

Simplify these fractions:

1. $\dfrac{24x^3y^6z^3}{8x^2y^2z} = 3xy^4z^2$

2. $\dfrac{3x^2 - 14xy - 5y^2}{x^2 - 25y^2} = \dfrac{(3x+y)(x-5y)}{(x+5y)(x-5y)} = \dfrac{3x+y}{x+5y}$

3. Re-write this fraction with a denominator of $(x+3)(x-5)(x+4)$.

$$\dfrac{x+2}{x^2+7x+12} = \dfrac{x+2}{(x+3)(x+4)} = \dfrac{(x+2)(x-5)}{(x+3)(x+4)(x-5)}$$

Try these:

1. $\dfrac{72x^4y^9z^{10}}{8x^3y^9z^5}$

2. $\dfrac{3x^2 - 13xy - 10y^2}{x^3 - 125y^3}$

3. Re-write this fraction with a denominator of $(x+2)(x+3)(x-7)$.

$$\dfrac{x+5}{x^2 - 5x - 14}$$

Simplify a radical.

To simplify a radical, follow these steps:

First factor the number or coefficient completely.

- For square roots, group like-factors in groups of 2. For cube roots, group like-factors in groups of 3. For n^{th} roots, group like-factors in groups of n.

- Now, for each of those groups, put one of that number outside the radical. Multiply these numbers by any number already in front of the radical. Any factors that were not combined in groups should be multiplied back together and left inside the radical.

- The index number of a radical is the little number on the front of the radical. For a cube root, the index is a 3. If no index appears, then the index is a 2 (for square roots).

- For variables inside the radical, divide the index number of the radical into each exponent. The quotient (the answer to the division) is the new exponent to be written on the variable outside the radical. The remainder from the division is the new exponent on the variable remaining inside the radical sign. If the remainder is zero, then the variable no longer appears inside the radical sign.

- Note: Remember that the square root of a negative number can be done by replacing the negative sign inside the square root sign with an "i" in front of the radical (to indicate an imaginary number). Then simplify the remaining positive radical by the normal method. Include the "i" outside the radical as part of the answer.

$$\sqrt{-18} = i\sqrt{18} = i\sqrt{3 \cdot 3 \cdot 2} = 3i\sqrt{2}$$

- Remember that if the index number is an odd number, you can still simplify the radical to get a negative solution.

Simplify:

1. $\sqrt{50a^4b^7} = \sqrt{5 \cdot 5 \cdot 2 \cdot a^4 b^7} = 5a^2 b^3 \sqrt{2b}$
2. $7x\sqrt[3]{16x^5} = 7x\sqrt[3]{2 \cdot 2 \cdot 2 \cdot 2 \cdot x^5} = 7x \cdot 2x\sqrt[3]{2x^2} = 14x^2 \sqrt[3]{2x^2}$

Try These :

1. $\sqrt{72a^9}$
2. $\sqrt{-98}$
3. $\sqrt[3]{-8x^6}$
4. $2x^3 y \sqrt[4]{243x^6 y^{11}}$

Rewrite expressions involving radicals as expressions with rational number exponents.

An expression with a radical sign can be rewritten using a rational exponent. The radicand becomes the base which will have the rational exponent. The index number on the front of the radical sign becomes the denominator of the rational exponent. The numerator of the rational exponent is the exponent which was originally inside the radical sign on the original base. Note: If no index number appears on the front of the radical, then it is a 2. If no exponent appears inside the radical, then use a 1 as the numerator of the rational exponent.

$$\sqrt[5]{b^3} = b^{3/5}$$
$$\sqrt[4]{ab^3} = a^{1/4} b^{3/4}$$

When an expression has a rational exponent, it can be rewritten using a radical sign. The denominator of the rational exponent becomes the index number on the front of the radical sign. The base of the original expression goes inside the radical sign. The numerator of the rational exponent is an exponent which can be placed either inside the radical sign on the original base or outside the radical as an exponent on the radical expression.

$$a^{2/9}b^{4/9}c^{8/9} = \sqrt[9]{a^2 b^4 c^8}$$
$$3^{1/5} = \sqrt[5]{3}$$

If an expression contains rational expressions with different denominators, rewrite the exponents with a common denominator and then change the problem into a radical.

$$a^{2/3}b^{1/2}c^{3/5} = a^{20/30}b^{15/30}c^{18/30} = \sqrt[30]{a^{20}b^{15}c^{18}}$$

TEACHER CERTIFICATION STUDY GUIDE

SKILL 1.8 Perform the four basic operations on rational and radical expressions.

The Order of Operations are to be followed when evaluating algebraic expressions. Follow these steps in order:

1. Simplify inside grouping characters such as parentheses, brackets, square root, fraction bar, etc.

2. Multiply out expressions with exponents.

3. Do multiplication or division, from left to right.

4. Do addition or subtraction, from left to right.

Samples of simplifying expressions with exponents:

$(^-2)^3 = -8$ $^-2^3 = {^-8}$
$(^-2)^4 = 16$ $^-2^4 = 16$ Note change of sign.
$(2/3)^3 = 8/27$
$5^0 = 1$
$4^{-1} = 1/4$

In order **to add or subtract** rational expressions, they must have a common denominator. If they don't have a common denominator, then factor the denominators to determine what factors are missing from each denominator to make the LCD. Multiply both numerator and denominator by the missing factor(s). Once the fractions have a common denominator, add or subtract their numerators, but keep the common denominator the same. Factor the numerator if possible and reduce if there are any factors that can be cancelled.

In order **to multiply** rational expressions, they do not have to have a common denominator. If you factor each numerator and denominator, you can cancel out any factor that occurs in both the numerator and denominator. Then multiply the remaining factors of the numerator together. Last multiply the remaining factors of the denominator together.

In order **to divide** rational expressions, the problem must be rewritten as the first fraction multiplied times the inverse of the second fraction. Once the problem has been written as a multiplication, factor each numerator and denominator.

MATHEMATICS 6-12

TEACHER CERTIFICATION STUDY GUIDE

Cancel out any factor that occurs in both the numerator and denominator. Then multiply the remaining factors of the numerator together. Last multiply the remaining factors of the denominator together.

1. $\dfrac{5}{x^2-9} - \dfrac{2}{x^2+4x+3} = \dfrac{5}{(x-3)(x+3)} - \dfrac{2}{(x+3)(x+1)} =$

$\dfrac{5(x+1)}{(x+1)(x-3)(x+3)} - \dfrac{2(x-3)}{(x+3)(x+1)(x-3)} = \dfrac{3x+11}{(x-3)(x+3)(x+1)}$

2. $\dfrac{x^2-2x-24}{x^2+6x+8} \times \dfrac{x^2+3x+2}{x^2-13x+42} = \dfrac{(x-6)(x+4)}{(x+4)(x+2)} \times \dfrac{(x+2)(x+1)}{(x-7)(x-6)} = \dfrac{x+1}{x-7}$

Try these:

1. $\dfrac{6}{x^2-1} + \dfrac{8}{x^2+7x+6}$

2. $\dfrac{x^2-9}{x^2-4} \div \dfrac{x^2+8x+15}{x^3+8}$

Add, subtract, multiply, divide, and simplify radical expressions.

Before you can add or subtract square roots, the numbers or expressions inside the radicals must be the same. First, simplify the radicals, if possible. If the numbers or expressions inside the radicals are the same, add or subtract the numbers (or like expressions) in front of the radicals. Keep the expression inside the radical the same. Be sure that the radicals are as simplified as possible.

Note: If the expressions inside the radicals are not the same, and can not be simplified to become the same, then they can not be combined by addition or subtraction.

To multiply 2 square roots together, follow these steps:

1. Multiply what is outside the radicals together.

2. Multiply what is inside the radicals together.

3. Simplify the radical if possible. Multiply whatever is in front of the radical times the expression that is coming out of the radical.

MATHEMATICS 6-12

TEACHER CERTIFICATION STUDY GUIDE

To divide one square root by another, follow these steps:

1. Work separately on what is inside or outside the square root sign.
2. Divide or reduce the coefficients outside the radical.
3. Divide any like variables outside the radical.
4. Divide or reduce the coefficients inside the radical.
5. Divide any like variables inside the radical.
6. If there is still a radical in the denominator, multiply both the numerator and denominator by the radical in the denominator. Simplify both resulting radicals and reduce again outside the radical (if possible).

Simplify:

1. $6\sqrt{7} + 2\sqrt{5} + 3\sqrt{7} = 9\sqrt{7} + 2\sqrt{5}$ These cannot be combined further.

2. $5\sqrt{12} + \sqrt{48} - 2\sqrt{75} = 5\sqrt{2 \cdot 2 \cdot 3} + \sqrt{2 \cdot 2 \cdot 2 \cdot 2 \cdot 3} - 2\sqrt{3 \cdot 5 \cdot 5} =$
$5 \cdot 2\sqrt{3} + 2 \cdot 2\sqrt{3} - 2 \cdot 5\sqrt{3} = 10\sqrt{3} + 4\sqrt{3} - 10\sqrt{3} \qquad = 4\sqrt{3}$

3. $(6\sqrt{15x})(7\sqrt{10x}) = 42\sqrt{150x^2} = 42\sqrt{2 \cdot 3 \cdot 5 \cdot 5 \cdot x^2} = 42 \cdot 5x\sqrt{2 \cdot 3} = 210x\sqrt{6}$

4. $\dfrac{105x^8 \sqrt{18x^5y^6}}{30x^2 \sqrt{27x^2y^4}} = \dfrac{7x^6(x^2)(y^3)\sqrt{2x}}{2(x)(y^2)\sqrt{3}} = \dfrac{7x^7y\sqrt{2x}}{2\sqrt{3}}$

$= \dfrac{7x^7y\sqrt{2x}}{2\sqrt{3}} \cdot \dfrac{\sqrt{3}}{\sqrt{3}} = \dfrac{7x^7y\sqrt{6x}}{6}$

Try these:

1. $6\sqrt{24} + 3\sqrt{54} - \sqrt{96}$

2. $(2x^2y\sqrt{18x})(7xy^7 \sqrt{4x})$

3. $\dfrac{125a^5 \sqrt{56a^4b^7}}{40a^2 \sqrt{40a^2b^8}}$

4. $2\sqrt{3} + 4\sqrt{5} + 6\sqrt{25} - 7\sqrt{9} + 2\sqrt{5} - 8\sqrt{20} - 6\sqrt{16} - 7\sqrt{3}$

SKILL 1.9 Solve equations containing radicals.

To solve a radical equation, follow these steps:

1. Get a radical alone on one side of the equation.
2. Raise both **sides** of the equation to the power equal to the index number. **Do not raise them to that power term by term, but raise the entire side to that power**. Combine any like terms.
3. If there is another radical still in the equation, repeat steps one and two (i.e. get that radical alone on one side of the equation and raise both sides to a power equal to the index). Repeat as necessary until the radicals are all gone.
4. Solve the resulting equation.
5. Once you have found the answer(s), substitute them back into the original equation to check them. Sometimes there are solutions that do not check in the original equation. These are extraneous solutions, which are not correct and must be eliminated. If a problem has more than one potential solution, each solution must be checked separately.

> NOTE: What this really means is that you can substitute the answers from any multiple choice test back into the question to determine which answer choice is correct.

Solve and **check**.

1. $\sqrt{2x+1} + 7 = x$

    $\sqrt{2x+1} = x - 7$

    $(\sqrt{2x+1})^2 = (x-7)^2$ ← BOTH sides are squared.

    $2x + 1 = x^2 - 14x + 49$

    $0 = x^2 - 16x + 48$

    $0 = (x - 12)(x - 4)$

    $x = 12, \ x = 4$

When you check these answers in the original equation, 12 checks; however, **4 does not check in the original equation**. Therefore, the only answer is $x = 12$.

2. $\sqrt{3x+4} = 2\sqrt{x-4}$

   $\left(\sqrt{3x+4}\right)^2 = \left(2\sqrt{x-4}\right)^2$

   $3x+4 = 4(x-4)$

   $3x+4 = 4x-16$

   $20 = x$ ← This checks in the original equaion.

3. $\sqrt[4]{7x-3} = 3$

   $\left(\sqrt[4]{7x-3}\right)^4 = 3^4$

   $7x-3 = 81$

   $7x = 84$

   $x = 12$ ← This checks out with the original equation.

4. $\sqrt{x} = {}^-3$

   $\left(\sqrt{x}\right)^2 = \left({}^-3\right)^2$

   $x = 9$ ← This does NOT check in the original equation. Since there is no other answer to check, the correct answer is the empty set or the null set or $\varnothing$.

Try these:

Solve and check.

1. $\sqrt{8x-24} + 14 = 2x$
2. $\sqrt{6x-2} = 5\sqrt{x-13}$

TEACHER CERTIFICATION STUDY GUIDE

SKILL 1.10 Multiply or divide binomials containing radicals.

The conjugate of a binomial is the same expression as the original binomial with the sign between the 2 terms changed.

$$\text{The conjugate of } 3+2\sqrt{5} \text{ is } 3-2\sqrt{5}.$$
$$\text{The conjugate of } \sqrt{5}-\sqrt{7} \text{ is } \sqrt{5}+\sqrt{7}.$$
$$\text{The conjugate of } {}^-6-\sqrt{11} \text{ is } {}^-6+\sqrt{11}.$$

To multiply binomials including radicals, "FOIL" the binomials together. (that is, distribute each term of the first binomial times each term of the second binomial). Multiply what is in front of the radicals together. Multiply what is inside of the two radicals together. Check to see if any of the radicals can be simplified. Combine like terms, if possible.

When one binomial is divided by another binomial, multiply both the numerator and denominator by the conjugate of the denominator. "FOIL" or distribute one binomial through the other binomial. Simplify the radicals, if possible, and combine like terms. Reduce the resulting fraction if every term is divisible outside the radical signs by the same number.

1. $(5+\sqrt{10})(4-3\sqrt{2}) = 20-15\sqrt{2}+4\sqrt{10}-3\sqrt{20} =$
   $20-15\sqrt{2}+4\sqrt{10}-6\sqrt{5}$

2. $(\sqrt{6}+5\sqrt{2})(3\sqrt{6}-8\sqrt{2}) = 3\sqrt{36}-8\sqrt{12}+15\sqrt{12}-40\sqrt{4} =$
   $3 \cdot 6 - 8 \cdot 2\sqrt{3} + 15 \cdot 2\sqrt{3} - 40 \cdot 2 = 18 - 16\sqrt{3} + 30\sqrt{3} - 80 = {}^-62 + 14\sqrt{3}$

3. $\dfrac{1-\sqrt{2}}{3+5\sqrt{2}} = \dfrac{1-\sqrt{2}}{3+5\sqrt{2}} \cdot \dfrac{3-5\sqrt{2}}{3-5\sqrt{2}} = \dfrac{3-5\sqrt{2}-3\sqrt{2}+5\sqrt{4}}{9-25\sqrt{4}} = \dfrac{3-5\sqrt{2}-3\sqrt{2}+10}{9-50} =$
   $\dfrac{13-8\sqrt{2}}{{}^-41}$ or $-\dfrac{13-8\sqrt{2}}{41}$ or $\dfrac{{}^-13+8\sqrt{2}}{41}$

Try These:
1. $(3+2\sqrt{6})(4-\sqrt{6})$
2. $(\sqrt{5}+2\sqrt{15})(\sqrt{3}-\sqrt{15})$
3. $\dfrac{6+2\sqrt{3}}{4-\sqrt{6}}$

SKILL 1.11 Solve quadratic equations by factoring, graphing, completing the square, or using the quadratic formula, including complex solutions.

A **quadratic equation** is written in the form $ax^2 + bx + c = 0$. To solve a quadratic equation by **factoring**, at least one of the factors must equal zero.

Example:
Solve the equation.

$x^2 + 10x - 24 = 0$

$(x+12)(x-2) = 0$ Factor.

$x + 12 = 0$ or $x - 2 = 0$ Set each factor equal to 0.

$x = {}^-12$ $x = 2$ Solve.

Check:

$x^2 + 10x - 24 = 0$

$({}^-12)^2 + 10({}^-12) - 24 = 0$ $(2)^2 + 10(2) - 24 = 0$

$144 - 120 - 24 = 0$ $4 + 20 - 24 = 0$

$0 = 0$ $0 = 0$

A quadratic equation that cannot be solved by factoring can be solved by **completing the square**.

Example:
Solve the equation.

$x^2 - 6x + 8 = 0$

$x^2 - 6x = {}^-8$ Move the constant to the right side.

$x^2 - 6x + 9 = {}^-8 + 9$ Add the square of half the coefficient of x to both sides.

$(x - 3)^2 = 1$ Write the left side as a perfect square.

$x - 3 = \pm\sqrt{1}$ Take the square root of both sides.

$x - 3 = 1$ $x - 3 = {}^-1$ Solve.

$x = 4$ $x = 2$

Check:

$x^2 - 6x + 8 = 0$

$4^2 - 6(4) + 8 = 0$ $2^2 - 6(2) + 8 = 0$

$16 - 24 + 8 = 0$ $4 - 12 + 8 = 0$

$0 = 0$ $0 = 0$

The general technique for **graphing** quadratics is the same as for graphing linear equations. Graphing quadratic equations, however, results in a parabola instead of a straight line.

Example:

Graph $y = 3x^2 + x - 2$.

x	$y = 3x^2 + x - 2$
-2	8
-1	0
0	-2
1	2
2	12

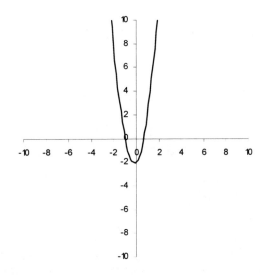

To solve a quadratic equation using the **quadratic formula**, be sure that your equation is in the form $ax^2 + bx + c = 0$. Substitute these values into the formula:

$$x = \frac{-b \pm \sqrt{b^2 - 4ac}}{2a}$$

Example:
Solve the equation.

$$3x^2 = 7 + 2x \rightarrow 3x^2 - 2x - 7 = 0$$

$a = 3 \quad b = {}^-2 \quad c = {}^-7$

$$x = \frac{-({}^-2) \pm \sqrt{({}^-2)^2 - 4(3)({}^-7)}}{2(3)}$$

$$x = \frac{2 \pm \sqrt{4 + 84}}{6}$$

$$x = \frac{2 \pm \sqrt{88}}{6}$$

$$x = \frac{2 \pm 2\sqrt{22}}{6}$$

$$x = \frac{1 \pm \sqrt{22}}{3}$$

To solve a quadratic equation(with x^2), rewrite the equation into the form:

$$ax^2 + bx + c = 0 \text{ or } y = ax^2 + bx + c$$

where a, b, and c are real numbers. Then substitute the values of a, b, and c into the quadratic formula:

$$x = \frac{-b \pm \sqrt{b^2 - 4ac}}{2a}$$

Simplify the result to find the answers. (Remember, there could be 2 real answers, one real answer, or 2 complex answers that include "i").

SKILL 1.12 Solve problems using quadratic equations.

Some word problems will give a quadratic equation to be solved. When the quadratic equation is found, set it equal to zero and solve the equation by factoring or the quadratic formula. Examples of this type of problem follow.

Example:

Alberta (A) is a certain distance north of Boston (B). The distance from Boston east to Carlisle (C) is 5 miles more than the distance from Alberta to Boston. The distance from Alberta to Carlisle is 10 miles more than the distance from Alberta to Boston. How far is Alberta from Carlisle?

Solution:

Since north and east form a right angle, these distances are the lengths of the legs of a right triangle. If the distance from Alberta to Boston is x, then from Boston to Carlisle is $x+5$, and the distance from Alberta to Carlisle is $x+10$.

The equation is: $AB^2 + BC^2 = AC^2$

$$x^2 + (x+5)^2 = (x+10)^2$$
$$x^2 + x^2 + 10x + 25 = x^2 + 20x + 100$$
$$2x^2 + 10x + 25 = x^2 + 20x + 100$$
$$x^2 - 10x - 75 = 0$$
$$(x-15)(x+5) = 0 \quad \text{Distance cannot be negative.}$$
$$x = 15 \quad \text{Distance from Alberta to Boston.}$$
$$x + 5 = 20 \quad \text{Distance from Boston to Carlisle.}$$
$$x + 10 = 25 \quad \text{Distance from Alberta to Carlisle.}$$

Example:

The square of a number is equal to 6 more than the original number. Find the original number.
Solution: If x = original number, then the equation is:

$$x^2 = 6 + x \quad \text{Set this equal to zero.}$$
$$x^2 - x - 6 = 0 \quad \text{Now factor.}$$
$$(x-3)(x+2) = 0$$
$$x = 3 \text{ or } x = {}^-2 \quad \text{Both solutions check out.}$$

Try these:

1. One side of a right triangle is 1 less than twice the shortest side, while the third side of the triangle is 1 more than twice the shortest side. Find all 3 sides.

2. Twice the square of a number equals 2 less than 5 times the number. Find the number(s).

Some word problems can be solved by setting up a quadratic equation or inequality. Examples of this type could be problems that deal with finding a maximum area. Examples follow:

Example:
A family wants to enclose 3 sides of a rectangular garden with 200 feet of fence. In order to have a garden with an area of **at least** 4800 square feet, find the dimensions of the garden. Assume that the fourth side of the garden is already bordered by a wall or a fence.

Solution:
Let x = distance from the wall

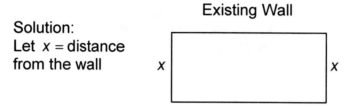

Then 2x feet of fence is used for these 2 sides. The remaining side of the garden would use the rest of the 200 feet of fence, that is, $200-2x$ feet of fence. Therefore the width of the garden is x feet and the length is $200-2x$ ft.

The area, $200x - 2x^2$, needs to be greater than or equal to 4800 sq. ft. So, this problem uses the inequality $4800 \leq 200x - 2x^2$. This becomes $2x^2 - 200x + 4800 \leq 0$. Solving this, we get:

$$200x - 2x^2 \geq 4800$$
$$-2x^2 + 200x - 4800 \geq 0$$
$$2\left(-x^2 + 100x - 2400\right) \geq 0$$
$$-x^2 + 100x - 2400 \geq 0$$
$$(-x + 60)(x - 40) \geq 0$$
$$-x + 60 \geq 0$$
$$-x \geq -60$$
$$x \leq 60$$
$$x - 40 \geq 0$$
$$x \geq 40$$

So the area will be at least 4800 square feet if the width of the garden is from 40 up to 60 feet. Quadratic equations can be used to model different real life situations. The graphs of these quadratics can be used to determine information about this real life situation.

SKILL 1.13 Use the discriminant to determine the nature of solutions of quadratic equations.

The discriminant of a quadratic equation is the part of the quadratic formula that is usually inside the radical sign, $b^2 - 4ac$.

$$x = \frac{-b \pm \sqrt{b^2 - 4ac}}{2a}$$

The radical sign is NOT part of the discriminant!! Determine the value of the discriminant by substituting the values of a, b, and c from $ax^2 + bx + c = 0$.

-If the value of the discriminant is **any negative number**, then there are **two complex roots** including "i".
-If the value of the discriminant is **zero**, then there is only **1 real rational root**. This would be a double root.
-If the value of the discriminant is **any positive number that is also a perfect square**, then there are **two real rational roots**. (There are no longer any radical signs.)
-If the value of the discriminant is **any positive number that is NOT a perfect square**, then there are **two real irrational roots**. (There are still unsimplified radical signs.)

Example:

Find the value of the discriminant for the following equations. Then determine the number and nature of the solutions of that quadratic equation.

$2x^2 - 5x + 6 = 0$
$a = 2$, $b = {}^-5$, $c = 6$ so $b^2 - 4ac = ({}^-5)^2 - 4(2)(6) = 25 - 48 = {}^-23$.

Since $^-23$ is a negative number, there are **two complex roots** including "i".

$$x = \frac{5}{4} + \frac{i\sqrt{23}}{4}, x = \frac{5}{4} - \frac{i\sqrt{23}}{4}$$

$3x^2 - 12x + 12 = 0$
$a = 3$, $b = {}^-12$, $c = 12$ so $b^2 - 4ac = ({}^-12)^2 - 4(3)(12) = 144 - 144 = 0$

Since 0 is the value of the discriminant, there is only
1 real rational root.
$$x = 2.$$

$6x^2 - x - 2 = 0$
$a = 6$, $b = {}^-1$, $c = {}^-2$ so $b^2 - 4ac = ({}^-1)^2 - 4(6)({}^-2) = 1 + 48 = 49$.

Since 49 is positive and is also a perfect square ($\sqrt{49} = 7$), then there are **two real rational roots**.

$$x = \frac{2}{3}, x = -\frac{1}{2}$$

Try these:
1. $6x^2 - 7x - 8 = 0$
2. $10x^2 - x - 2 = 0$
3. $25x^2 - 80x + 64 = 0$

SKILL 1.14 Determine a quadratic equation from known roots.

Follow these steps to write a quadratic equation from its roots:

1. Add the roots together. The answer is their **sum**. Multiply the roots together. The answer is their **product**.

2. A quadratic equation can be written using the sum and product like this:

$$x^2 + \text{(opposite of the sum)}x + \text{product} = 0$$

3. If there are any fractions in the equation, multiply every term by the common denominator to eliminate the fractions. This is the quadratic equation.

4. If a quadratic equation has only 1 root, use it twice and follow the first 3 steps above.

Example:
Find a quadratic equation with roots of 4 and $^-9$.

Solutions:
The sum of 4 and $^-9$ is $^-5$. The product of 4 and $^-9$ is $^-36$.
The equation would be:

$$x^2 + \text{(opposite of the sum)}x + \text{product} = 0$$
$$x^2 + 5x - 36 = 0$$

Find a quadratic equation with roots of $5+2i$ and $5-2i$.

Solutions:
The sum of $5+2i$ and $5-2i$ is 10. The product of $5+2i$ and $5-2i$ is

$25 - 4i^2 = 25 + 4 = 29$.

The equation would be:

$$x^2 + (\text{opposite of the sum})x + \text{product} = 0$$
$$x^2 - 10x + 29 = 0$$

Find a quadratic equation with roots of $2/3$ and $^-3/4$.

Solutions:
The sum of $2/3$ and $^-3/4$ is $^-1/12$. The product of $2/3$ and $^-3/4$ is $^-1/2$.
The equation would be :

$$x^2 + (\text{opposite of the sum})x + \text{product} = 0$$
$$x^2 + 1/12\,x - 1/2 = 0$$

Common denominator = 12, so multiply by 12.

$$12(x^2 + 1/12\,x - 1/2 = 0$$
$$12x^2 + 1x - 6 = 0$$
$$12x^2 + x - 6 = 0$$

Try these:

1. Find a quadratic equation with a root of 5.
2. Find a quadratic equation with roots of $8/5$ and $^-6/5$.
3. Find a quadratic equation with roots of 12 and $^-3$.

SKILL 1.15 Identify the graphs of quadratic inequalities.

To graph an inequality, graph the quadratic as if it was an equation; however, if the inequality has just a $>$ or $<$ sign, then make the curve itself dotted. Shade above the curve for $>$ or $\geq$. Shade below the curve for $<$ or $\leq$.

Examples:

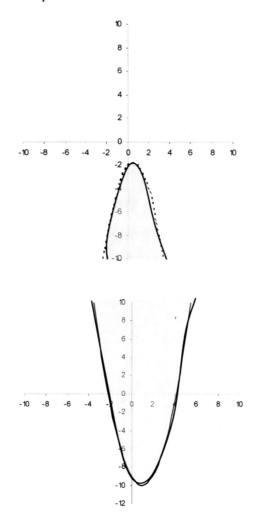

TEACHER CERTIFICATION STUDY GUIDE

SKILL 1.16 Solve real-world problems using direct and inverse variations.

-If two things vary directly, as one gets larger, the other also gets larger. If one gets smaller, then the other gets smaller too. If x and y vary directly, there should be a constant, c, such that $y = cx$. Something can also vary directly with the square of something else, $y = cx^2$.

-If two things vary inversely, as one gets larger, the other one gets smaller instead. If x and y vary inversely, there should be a constant, c, such that $xy = c$ or $y = c/x$. Something can also vary inversely with the square of something else, $y = c/x^2$.

Example: If $30 is paid for 5 hours work, how much would be paid for 19 hours work?

This is direct variation and $30 = 5c, so the constant is 6 ($6/hour). So $y = 6(19)$ or $y = \$114$.

This could also be done as a proportion:

$$\frac{\$30}{5} = \frac{y}{19}$$

$$5y = 570$$
$$y = 114$$

Example: On a 546 mile trip from Miami to Charlotte, one car drove 65 mph while another car drove 70 mph. How does this affect the driving time for the trip?
This is an inverse variation, since increasing your speed should decrease your driving time. Using the equation: rate × time = distance, rt = d.

$$65t = 546 \quad \text{and} \quad 70t = 546$$
$$t = 8.4 \quad \text{and} \quad t = 7.8$$

slower speed, more time   faster speed, less time

MATHEMATICS 6-12

Example: A 14" pizza from Azzip Pizza costs $8.00. How much would a 20" pizza cost if its price was based on the same price per square inch?

Here the price is directly proportional to the square of the radius. Using a proportion:

$$\frac{\$8.00}{7^2 \pi} = \frac{x}{10^2 \pi}$$

$$\frac{8}{153.86} = \frac{x}{314}$$

$$16.33 = x$$

$16.33 would be the price of the large pizza.

SKILL 1.17 Solve systems of linear equations or inequalities.

Word problems can sometimes be solved by using a system of two equations with 2 unknowns. This system can then be solved using **substitution**, the **addition-subtraction method**, or **determinants**.

Example: Farmer Greenjeans bought 4 cows and 6 sheep for $1700. Mr. Ziffel bought 3 cows and 12 sheep for $2400. If all the cows were the same price and all the sheep were another price, find the price charged for a cow or for a sheep.

Let x = price of a cow
Let y = price of a sheep

Then Farmer Greenjeans' equation would be: $4x + 6y = 1700$
Mr. Ziffel's equation would be: $3x + 12y = 2400$

To solve by **addition-subtraction**:

Multiply the first equation by $^-2$: $^-2(4x + 6y = 1700)$
Keep the other equation the same : $(3x + 12y = 2400)$
By doing this, the equations can be added to each other to eliminate one variable and solve for the other variable.

$$^-8x - 12y = ^-3400$$
$$3x + 12y = 2400 \quad \text{Add these equations.}$$
$$^-5x \quad\quad = ^-1000$$

$x = 200 \leftarrow$ the price of a cow was $200.
Solving for y, $y = 150 \leftarrow$ the price of a sheep, $150.

To solve by **substitution**:

Solve one of the equations for a variable. (Try to make an equation without fractions if possible.) Substitute this expression into the equation that you have not yet used. Solve the resulting equation for the value of the remaining variable.

$$4x + 6y = 1700$$
$$3x + 12y = 2400 \leftarrow \text{Solve this equation for } x.$$

It becomes $x = 800 - 4y$. Now substitute $800 - 4y$ in place of x in the OTHER equation. $4x + 6y = 1700$ now becomes:

$$4(800 - 4y) + 6y = 1700$$
$$3200 - 16y + 6y = 1700$$
$$3200 - 10y = 1700$$
$$^-10y = {}^-1500$$
$$y = 150, \text{ or } \$150 \text{ for a sheep.}$$

Substituting 150 back into an equation for y, find x.
$$4x + 6(150) = 1700$$
$$4x + 900 = 1700$$
$$4x = 800 \text{ so } x = 200 \text{ for a cow.}$$

To solve by **determinants**:

Let x = price of a cow
Let y = price of a sheep

Then Farmer Greenjeans' equation would be: $4x + 6y = 1700$
Mr. Ziffel's equation would be: $3x + 12y = 2400$

To solve this system using determinants, make one 2 by 2 determinant divided by another 2 by 2 determinant. The bottom determinant is filled with the x and y term coefficients. The top determinant is almost the same as this bottom determinant. The only difference is that when you are solving for x, the x coefficients are replaced with the constants. Likewise, when you are solving for y, the y coefficients are replaced with the constants. To find the value of a 2 by 2 determinant, $\begin{pmatrix} a & b \\ c & d \end{pmatrix}$, is found by $ad - bc$.

$$x = \frac{\begin{pmatrix} 1700 & 6 \\ 2400 & 12 \end{pmatrix}}{\begin{pmatrix} 4 & 6 \\ 3 & 12 \end{pmatrix}} = \frac{1700(12) - 6(2400)}{4(12) - 6(3)} = \frac{20400 - 14400}{48 - 18} = \frac{6000}{30} = 200$$

$$y = \frac{\begin{pmatrix} 4 & 1700 \\ 3 & 2400 \end{pmatrix}}{\begin{pmatrix} 4 & 6 \\ 3 & 12 \end{pmatrix}} = \frac{2400(4) - 3(1700)}{4(12) - 6(3)} = \frac{9600 - 5100}{48 - 18} = \frac{4500}{30} = 150$$

NOTE: The bottom determinant is always the same value for each letter.

Word problems can sometimes be solved by using a system of three equations with 3 unknowns. This system can then be solved using **substitution**, the **addition-subtraction method**, or **determinants**.

To solve by **substitution**:

Example: Mrs. Allison bought 1 pound of potato chips, a 2 pound beef roast, and 3 pounds of apples for a total of $ 8.19. Mr. Bromberg bought a 3 pound beef roast and 2 pounds of apples for $ 9.05. Kathleen Kaufman bought 2 pounds of potato chips, a 3 pound beef roast, and 5 pounds of apples for $ 13.25. Find the per pound price of each item.

Let x = price of a pound of potato chips
Let y = price of a pound of roast beef
Let z = price of a pound of apples

Mrs. Allison's equation would be: $1x + 2y + 3z = 8.19$
Mr. Bromberg's equation would be: $3y + 2z = 9.05$
K. Kaufman's equation would be: $2x + 3y + 5z = 13.25$

Take the first equation and solve it for x. (This was chosen because x is the easiest variable to get alone in this set of equations.) This equation would become:

$$x = 8.19 - 2y - 3z$$

Substitute this expression into the other equations in place of the letter x:

$$3y + 2z = 9.05 \leftarrow \text{equation 2}$$
$$2(8.19 - 2y - 3z) + 3y + 5z = 13.25 \leftarrow \text{equation 3}$$

Simplify the equation by combining like terms:

$$3y + 2z = 9.05 \leftarrow \text{equation 2}$$
$$\text{*} \; {}^-1y - 1z = {}^-3.13 \leftarrow \text{equation 3}$$

Solve equation 3 for either y or z:

$y = 3.13 - z$ Substitute this into equation 2 for y:

$$3(3.13 - z) + 2z = 9.05 \leftarrow \text{equation 2}$$
$${}^-1y - 1z = {}^-3.13 \leftarrow \text{equation 3}$$

Combine like terms in equation 2:

$$9.39 - 3z + 2z = 9.05$$
$$z = .34 \quad \text{per pound price of apples}$$

Substitute .34 for z in the starred equation above to solve for y:

$y = 3.13 - z$ becomes $y = 3.13 - .34$, so
$y = 2.79 = $ per pound price of roast beef

Substituting .34 for z and 2.79 for y in one of the original equations, solve for x:

$$1x + 2y + 3z = 8.19$$
$$1x + 2(2.79) + 3(.34) = 8.19$$
$$x + 5.58 + 1.02 = 8.19$$
$$x + 6.60 = 8.19$$
$$x = 1.59 \quad \text{per pound of potato chips}$$

$(x, y, z) = (1.59, 2.79, .34)$

To solve by **addition-subtraction**:

Choose a letter to eliminate. Since the second equation is already missing an x, let's eliminate x from equations 1 and 3.

1) $1x + 2y + 3x = 8.19$ ← Multiply by $^-2$ below.
2) $3y + 2z = 9.05$
3) $2x + 3y + 5z = 13.25$

$^-2(1x + 2y + 3z = 8.19)\quad =\quad ^-2x - 4y - 6z = ^-16.38$
Keep equation 3 the same : $\quad 2x + 3y + 5z = 13.25$

By doing this, the equations $\quad\quad\quad ^-y - z = ^-3.13$ ← equation 4
can be added to each other to
eliminate one variable.

The equations left to solve are equations 2 and 4:
$\quad ^-y - z = ^-3.13$ ← equation 4
$\quad 3y + 2z = 9.05$ ← equation 2

Multiply equation 4 by 3: $\quad 3(^-y - z = ^-3.13)$
Keep equation 2 the same: $\quad 3y + 2z = 9.05$

$\quad\quad ^-3y - 3z = ^-9.39$
$\quad\quad \underline{3y + 2z = 9.05}\quad\quad$ Add these equations.
$\quad\quad\quad ^-1z = ^-.34$
$\quad\quad\quad\quad z = .34$ ← the per pound price of apples
solving for y, $y = 2.79$ ← the per pound roast beef price
solving for x, $x = 1.59$ ← potato chips, per pound price

To solve by **substitution**:

Solve one of the 3 equations for a variable. (Try to make an equation without fractions if possible.) Substitute this expression into the other 2 equations that you have not yet used.

1) $1x + 2y + 3z = 8.19$ ← Solve for x.
2) $3y + 2z = 9.05$
3) $2x + 3y + 5z = 13.25$

Equation 1 becomes $x = 8.19 - 2y - 3z$.

Substituting this into equations 2 and 3, they become:
2) $3y + 2z = 9.05$
3) $2(8.19 - 2y - 3z) + 3y + 5z = 13.25$
   $16.38 - 4y - 6z + 3y + 5z = 13.25$
   $^-y - z = {}^-3.13$

The equations left to solve are :

$3y + 2z = 9.05$

$^-y - z = {}^-3.13$ ← Solve for either y or z.

It becomes $y = 3.13 - z$. Now substitute $3.13 - z$ in place of y in the OTHER equation. $3y + 2z = 9.05$ now becomes:

$$3(3.13 - z) + 2z = 9.05$$
$$9.39 - 3z + 2z = 9.05$$
$$9.39 - z = 9.05$$
$$^-z = {}^-.34$$
$$z = .34 \text{, or \$.34/lb of apples}$$

Substituting .34 back into an equation for z, find y.
$3y + 2z = 9.05$
$3y + 2(.34) = 9.05$
$3y + .68 = 9.05$ so $y = 2.79$/lb of roast beef

Substituting .34 for z and 2.79 for y into one of the original equations, it becomes:
$2x + 3y + 5z = 13.25$
$2x + 3(2.79) + 5(.34) = 13.25$
$2x + 8.37 + 1.70 = 13.25$
$2x + 10.07 = 13.25$, so $x = 1.59$/lb of potato chips

To solve by **determinants**:

Let x = price of a pound of potato chips
Let y = price of a pound of roast beef
Let z = price of a pound of apples

1) $1x + 2y + 3z = 8.19$
2) $3y + 2z = 9.05$
3) $2x + 3y + 5z = 13.25$

To solve this system using determinants, make one 3 by 3 determinant divided by another 3 by 3 determinant. The bottom determinant is filled with the x, y, and z term coefficients. The top determinant is almost the same as this bottom determinant. The only difference is that when you are solving for x, the x coefficients are replaced with the constants. When you are solving for y, the y coefficients are replaced with the constants. Likewise, when you are solving for z, the z coefficients are replaced with the constants. To find the value of a 3 by 3 determinant,

$$\begin{pmatrix} a & b & c \\ d & e & f \\ g & h & i \end{pmatrix}$$ is found by the following steps:

Copy the first two columns to the right of the determinant:

$$\left(\begin{array}{ccc|cc} a & b & c & a & b \\ d & e & f & d & e \\ g & h & i & g & h \end{array}\right)$$

Multiply the diagonals from top left to bottom right, and add these diagonals together.

$$\left(\begin{array}{ccc|cc} a^* & b^\circ & c^\bullet & a & b \\ d & e^* & f^\circ & d^\bullet & e \\ g & h & i^* & g^\circ & h^\bullet \end{array}\right) = a^* e^* i^* + b^\circ f^\circ g^\circ + c^\bullet d^\bullet h^\bullet$$

Then multiply the diagonals from bottom left to top right, and add these diagonals together.

$$\begin{pmatrix} a & b & c^* \\ d & e^* & f^\circ \\ g^* & h^\circ & i^\bullet \end{pmatrix} \begin{matrix} a^\circ & b^\bullet \\ d^\bullet & e \\ g & h \end{matrix} = g^*e^*c^* + h^\circ f^\circ a^\circ + i^\bullet d^\bullet b^\bullet$$

Subtract: The first diagonal total minus the second diagonal total:

$$(= a^*e^*i^* + b^\circ f^\circ g^\circ + c^\bullet d^\bullet h^\bullet) - (= g^*e^*c^* + h^\circ f^\circ a^\circ + i^\bullet d^\bullet b^\bullet)$$

This gives the value of the determinant. To find the value of a variable, divide the value of the top determinant by the value of the bottom determinant.

1) $1x + 2y + 3z = 8.19$
2) $3y + 2z = 9.05$
3) $2x + 3y + 5z = 13.25$

$$x = \frac{\begin{pmatrix} 8.19 & 2 & 3 \\ 9.05 & 3 & 2 \\ 13.25 & 3 & 5 \end{pmatrix}}{\begin{pmatrix} 1 & 2 & 3 \\ 0 & 3 & 2 \\ 2 & 3 & 5 \end{pmatrix}}$$ solve each determinant using the method shown below

Multiply the diagonals from top left to bottom right, and add these diagonals together.

$$\begin{pmatrix} 8.19^* & 2^\circ & 3^\bullet \\ 9.05 & 3^* & 2^\circ \\ 13.25 & 3 & 5^* \end{pmatrix} \begin{matrix} 8.19 & 2 \\ 9.05^\bullet & 3 \\ 13.25^\circ & 3^\bullet \end{matrix}$$

$$= (8.19^*)(3^*)(5^*) + (2^\circ)(2^\circ)(13.25^\circ) + (3^\bullet)(9.05^\bullet)(3^\bullet)$$

Then multiply the diagonals from bottom left to top right, and add these diagonals together.

$$\begin{pmatrix} 8.19 & 2 & 3^* \\ 9.05 & 3^* & 2^\circ \\ 13.25^* & 3^\circ & 5^\bullet \end{pmatrix} \begin{matrix} 8.19^\circ & 2^\bullet \\ 9.05^\bullet & 3 \\ 13.25 & 3 \end{matrix}$$

$$= (13.25^*)(3^*)(3^*) + (3^\circ)(2^\circ)(8.19^\circ) + (5^\bullet)(9.05^\bullet)(2^\bullet)$$

Subtract the first diagonal total minus the second diagonal total:

$$(8.19^*)(3^*)(5^*) + (2^\circ)(2^\circ)(13.25^\circ) + (3^\bullet)(9.05^\bullet)(3^\bullet) = 257.30$$
$$- \underline{(13.25^*)(3^*)(3^*) + (3^\circ)(2^\circ)(8.19^\circ) + (5^\bullet)(9.05^\bullet)(2^\bullet) = ^- 258.89}$$
$$^-1.59$$

Use the same multiplying and subtraction procedure for the bottom determinant to get $^-1$ as an answer. Now divide:

$$\frac{^-1.59}{^-1} = \$1.59/\text{lb of potato chips}$$

$$y = \frac{\begin{pmatrix} 1 & 8.19 & 3 \\ 0 & 9.05 & 2 \\ 2 & 13.25 & 5 \end{pmatrix}}{\begin{pmatrix} 1 & 2 & 3 \\ 0 & 3 & 2 \\ 2 & 3 & 5 \end{pmatrix}} = \frac{^-2.79}{^-1} = \$2.79/\text{lb of roast beef}$$

NOTE: The bottom determinant is always the same value for each letter.

$$z = \frac{\begin{pmatrix} 1 & 2 & 8.19 \\ 0 & 3 & 9.05 \\ 2 & 3 & 13.25 \end{pmatrix}}{\begin{pmatrix} 1 & 2 & 3 \\ 0 & 3 & 2 \\ 2 & 3 & 5 \end{pmatrix}} = \frac{^-.34}{^-1} = \$.34/\text{lb of apples}$$

Skill 1.18 Solve systems of linear inequalities graphically.

To graph an inequality, solve the inequality for y. This gets the inequality in **slope intercept form**, (for example : $y < mx + b$). The point (0,b) is the y-intercept and m is the line's slope.

- If the inequality solves to $x \geq$ **any number**, then the graph includes a **vertical line**.

- If the inequality solves to $y \leq$ **any number**, then the graph includes a **horizontal line**.

- When graphing a linear inequality, the line will be dotted if the inequality sign is < or >. If the inequality signs are either ≥ or ≤, the line on the graph will be a solid line. Shade above the line when the inequality sign is ≥ or >. Shade below the line when the inequality sign is < or ≤. For inequalities of the forms $x >$ number, $x \leq$ number, $x <$ number, or $x \geq$ number, draw a vertical line (solid or dotted). Shade to the right for > or ≥. Shade to the left for < or ≤.

Remember: **Dividing or multiplying by a negative number will reverse the direction of the inequality sign.**

Use these rules to graph and shade each inequality. The solution to a system of linear inequalities consists of the part of the graph that is shaded for each inequality. For instance, if the graph of one inequality was shaded with red, and the graph of another inequality was shaded with blue, then the overlapping area would be shaded purple. The purple area would be the points in the solution set of this system.

Example: Solve by graphing:

$x + y \leq 6$

$x - 2y \leq 6$

Solving the inequalities for y, they become:

$y \leq {}^-x + 6$ (y intercept of 6 and slope = $^-1$)

$y \geq 1/2 x - 3$ (y intercept of $^-3$ and slope = $1/2$)

A graph with shading is shown below:

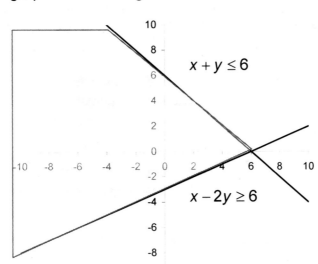

SKILL 1.19 Formulate or identify systems of linear equations or inequalities to solve real-world problems.

Some word problems can be solved using a system of equations or inequalities. Watch for words like greater than, less than, at least, or no more than which indicate the need for inequalities.

Example: The YMCA wants to sell raffle tickets to raise at least $32,000. If they must pay $7,250 in expenses and prizes out of the money collected from the tickets, how many tickets worth $25 each must they sell?

Solution: Since they want to raise **at least $32,000**, that means they would be happy to get 32,000 **or more**. This requires an inequality.

Let x = number of tickets sold
Then $25x$ = total money collected for x tickets

Total money minus expenses is greater than $32,000.

$$25x - 7250 \geq 32000$$
$$25x \geq 39250$$
$$x \geq 1570$$

If they sell **1,570 tickets or more**, they will raise AT LEAST $32,000.

MATHEMATICS 6-12

Example: The Simpsons went out for dinner. All 4 of them ordered the aardvark steak dinner. Bert paid for the 4 meals and included a tip of $12 for a total of $84.60. How much was an aardvark steak dinner?

Let x = the price of one aardvark dinner
So $4x$ = the price of 4 aardavark dinners

$4x = 84.60 - 12$

$4x = 72.60$

$x = \dfrac{72.60}{4} = \$18.15$ The price of one aardvark dinner.

Example: Farmer Greenjeans bought 4 cows and 6 sheep for $1700. Mr. Ziffel bought 3 cows and 12 sheep for $2400. If all the cows were the same price and all the sheep were another price, find the price charged for a cow or for a sheep.

Let x = price of a cow
Let y = price of a sheep

Then Farmer Greenjeans' equation would be: $4x + 6y = 1700$
Mr. Ziffel's equation would be: $3x + 12y = 2400$

To solve by **addition-subtraction**:

Multiply the first equation by $^-2$: $^-2(4x + 6y = 1700)$
Keep the other equation the same : $(3x + 12y = 2400)$
By doing this, the equations can be added to each other to eliminate one variable and solve for the other variable.

$^-8x - 12y = ^-3400$
$3x + 12y = 2400$ Add these equations.

$^-5x \quad\quad = ^-1000$

$x = 200$ ← the price of a cow was $200.
Solving for y, $y = 150$ ← the price of a sheep, $150.
(This problem can also be solved by substitution or determinants.)

Example: John has 9 coins, which are either dimes or nickels, that are worth $.65. Determine how many of each coin he has.

Let d = number of dimes.
Let n = number of nickels.
The number of coins total 9.
The value of the coins equals 65.

Then: $n + d = 9$
$5n + 10d = 65$

Multiplying the first equation by $^-5$, it becomes:

$^-5n - 5d = {}^-45$
$\underline{5n + 10d = 65}$
$5d = 20$

$d = 4$ There are 4 dimes, so there are $(9-4)$ or 5 nickels.

Example: Sharon's Bike Shoppe can assemble a 3 speed bike in 30 minutes or a 10 speed bike in 60 minutes. The profit on each bike sold is $60 for a 3 speed or $75 for a 10 speed bike. How many of each type of bike should they assemble during an 8 hour day (480 minutes) to make the maximum profit? Total daily profit must be at least $300.

Let x = number of 3 speed bikes.
y = number of 10 speed bikes.

Since there are only 480 minutes to use each day,

$30x + 60y \leq 480$ is the first inequality.

Since the total daily profit must be at least $300,

$60x + 75y \geq 300$ is the second inequality.

$30x + 65y \leq 480$ solves to $y \leq 8 - 1/2\,x$
$60y \leq -30x + 480$
$y \leq -\dfrac{1}{2}x + 8$

$60x + 75y \geq 300$ solves to $y \geq 4 - 4/5\, x$

$75y + 60x \geq 300$

$\quad 75y \geq -60x + 300$

$\quad y \geq -\dfrac{4}{5}x + 4$

Graph these 2 inequalities:

$\quad y \leq 8 - 1/2\, x$

$\quad y \geq 4 - 4/5\, x$

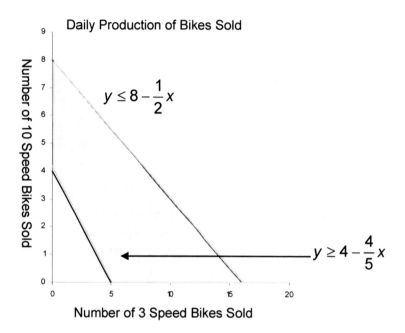

Daily Production of Bikes Sold

Realize that $x \geq 0$ and $y \geq 0$, since the number of bikes assembled can not be a negative number. Graph these as additional constraints on the problem. The number of bikes assembled must always be an integer value, so points within the shaded area of the graph must have integer values. The maximum profit will occur at or near a corner of the shaded portion of this graph. Those points occur at (0,4), (0,8), (16,0), or (5,0).

Since profits are $60/3-speed or $75/10-speed, the profit would be:

$\quad$ (0,4) $\quad 60(0) + 75(4) = 300$

$\quad$ (0,8) $\quad 60(0) + 75(8) = 600$

$\quad$ (16,0) $\quad 60(16) + 75(0) = 960$ ← Maximum profit

$\quad$ (5,0) $\quad 60(5) + 75(0) = 300$

The maximum profit would occur if 16 3-speed bikes are made daily.

TEACHER CERTIFICATION STUDY GUIDE

SKILL 1.20 Solve equations or inequalities involving absolute value.

To solve an absolute value equation, follow these steps:

1. Get the absolute value expression alone on one side of the equation.

2. Split the absolute value equation into 2 separate equations without absolute value bars. Write the expression inside the absolute value bars (without the bars) equal to the expression on the other side of the equation. Now write the expression inside the absolute value bars equal to the opposite of the expression on the other side of the equation.

3. Now solve each of these equations.

4. **Check each answer by substituting them into the original equation** (with the absolute value symbol). There will be answers that do not check in the original equation. These answers are discarded as they are **extraneous solutions**. If all answers are discarded as incorrect, then the answer to the equation is $\varnothing$, which means the empty set or the null set. (0, 1, or 2 solutions could be correct.)

To solve an absolute value inequality, follow these steps:

1. Get the absolute value expression alone on one side of the inequality.

   Remember: **Dividing or multiplying by a negative number will reverse the direction of the inequality sign.**

2. Remember what the inequality sign is at this point.

3. Split the absolute value inequality into 2 separate inequalities without absolute value bars. First rewrite the inequality without the absolute bars and solve it. Next write the expression inside the absolute value bar followed by the opposite inequality sign and then by the opposite of the expression on the other side of the inequality. Now solve it.

4. If the sign in the inequality on step 2 is $<$ or $\leq$, the answer is those 2 inequalities connected by the word **and**. The solution set consists of the points between the 2 numbers on the number line. If the sign in the inequality on step 2 is $>$ or $\geq$, the answer is those 2 inequalities connected by the word **or**. The solution set consists of the points outside the 2 numbers on the number line.

MATHEMATICS 6-12

If an expression inside an absolute value bar is compared to a negative number, the answer can also be either all real numbers or the empty set ($\emptyset$).

For instance,
$$|x+3| < {}^-6$$
would have the empty set as the answer, since an absolute value is always positive and will never be less than $^-6$. However,
$$|x+3| > {}^-6$$
would have all real numbers as the answer, since an absolute value is always positive or at least zero, and will never be less than -6. In similar fashion,
$$|x+3| = {}^-6$$
would never check because an absolute value will never give a negative value.

Example: Solve and check:

$$|2x - 5| + 1 = 12$$
$$|2x - 5| = 11 \quad \text{Get absolute value alone.}$$

Rewrite as 2 equations and solve separately.

same equation without absolute value		same equation without absolute value but right side is opposite
$2x - 5 = 11$		$2x - 5 = {}^-11$
$2x = 16$	and	$2x = {}^-6$
$x = 8$		$x = {}^-3$

Checks:

$\|2x - 5\| + 1 = 12$	$\|2x - 5\| + 1 = 12$
$\|2(8) - 5\| + 1 = 12$	$\|2({}^-3) - 5\| + 1 = 12$
$\|11\| + 1 = 12$	$\|{}^-11\| + 1 = 12$
$12 = 12$	$12 = 12$

This time both 8 and $^-3$ check.

Example: Solve and check:

$$2|x-7|-13 \geq 11$$
$$2|x-7| \geq 24 \quad \text{Get absolute value alone.}$$
$$|x-7| \geq 12$$

Rewrite as 2 inequalities and solve separately.

same inequality without absolute value		same inequality without absolute value but right side and inequality sign are both the opposite
$x - 7 \geq 12$	or	$x - 7 \leq {}^-12$
$x \geq 19$	or	$x \leq {}^-5$

SKILL 1.21 Expand given binomials to a specified positive integral power.

The binomial expansion theorem is another method used to find the coefficients of $(x + y)$. Although Pascal's Triangle is easy to use for small values of n, it can become cumbersome to use with larger values of n.

Binomial Theorem:

For any positive value of n,

$$(x+y)^n = x^n + \frac{n!}{(n-1)!1!}x^{n-1}y + \frac{n!}{(n-2)!2!}x^{n-2}y^2 + \frac{n!}{(n-3)!3!}x^{n-3}y^3 + \frac{n!}{1!(n-1)!}xy^{n-1} + y^n$$

Sample Problem:
1. Expand $(3x + y)^5$

$$(3x)^5 + \frac{5!}{4!1!}(3x)^4 y^1 + \frac{5!}{3!2!}(3x)^3 y^2 + \frac{5!}{2!3!}(3x)^2 y^3 + \frac{5!}{1!4!}(3x)^1 y^4 + y^5 =$$
$$243x^5 + 405x^4y + 270x^3y^2 + 90x^2y^3 + 15xy^4 + y^5$$

SKILL 1.22 Determine a specified term in the expansion of given binomials.

Any term of a binomial expansion can be written individually. For example, the y value of the seventh term of $(x+y)^n$, would be raised to the 6th power and since the sum of exponents on x and y must equal seven, then the x must be raised to the $n-6$ power. The formula to find the rth term of a binomial expansion is:

$$\frac{n!}{[n-(r-1)]!(r-1)!}x^{n-(r-1)}y^{r-1}$$

where $r =$ the number of the desired term and $n =$ the power of the binomial.

Sample Problem:

1. Find the third term of $(x+2y)^{11}$

$x^{n-(r-1)}$	y^{r-1}	Find x and y exponents.
$x^{11-(3-1)}$	y^{3-1}	
x^9	y^2	$y=2y$
$\frac{11!}{9!2!}(x^9)(2y)^2$		Substitute known values.
$220x^9y^2$		Solution.

Practice problems:

1. $(x+y)^7$; 5th term
2. $(3x-y)^9$; 3rd term

SKILL 1.23 Solve polynomial equations by factoring.

- To factor a polynomial, follow these steps:

a. **Factor out any GCF** (greatest common factor)

b. For a binomial (2 terms), check to see if the problem is the **difference of perfect squares**. If both factors are perfect squares, then it factors this way:
$$a^2 - b^2 = (a-b)(a+b)$$

If the problem is not the difference of perfect squares, then check to see if the problem is either the sum or difference of perfect cubes.

$$x^3 - 8y^3 = (x-2y)(x^2 + 2xy + 4y^2) \quad \leftarrow \text{difference}$$
$$64a^3 + 27b^3 = (4a+3b)(16a^3 - 12ab + 9b^2) \quad \leftarrow \text{sum}$$

** The sum of perfect squares does NOT factor.

c. Trinomials could be perfect squares. Trinomials can be factored into 2 binomials (un-FOILing). Be sure the terms of the trinomial are in descending order. If last sign of the trinomial is a "+", then the signs in the parentheses will be the same as the sign in front of the second term of the trinomial. If the last sign of the trinomial is a "-", then there will be one "+" and one "-" in the two parentheses. The first term of the trinomial can be factored to equal the first terms of the two factors. The last term of the trinomial can be factored to equal the last terms of the two factors. Work backwards to determine the correct factors to multiply together to get the correct center term.

Factor the following examples completely:

1. $4x^2 - 25y^2$
2. $6b^2 - 2b - 8$
3. Find a factor of $6x^2 - 5x - 4$

    a. $(3x+2)$ b. $(3x-2)$ c. $(6x-1)$ d. $(2x+1)$

Answers:

1. No GCF; this is the difference of perfect squares.

$$4x^2 - 25y^2 = (2x - 5y)(2x + 5y)$$

2. GCF of 2; Try to factor into 2 binomials:

$$6b^2 - 2b - 8 = 2(3b^2 - b - 4)$$

Signs are one "+", one "−". $3b^2$ factors into $3b$ and b. Find factors of 4: 1 & 4; 2 & 2.

$$6b^2 - 2b - 8 = 2(3b^2 - b - 4) = 2(3b - 4)(b + 1)$$

3. If an answer choice is correct, find the other factor:

    a. $(3x+2)(2x-2) = 6x^2 - 2x - 4$
    b. $(3x-2)(2x+2) = 6x^2 + 2x - 4$
    c. $(6x-1)(x+4) = 6x^2 + 23x - 4$
    d. $(2x+1)(3x-4) = 6x^2 - 5x - 4$ ← correct factors

SKILL 1.24 Perform vector addition, subtraction, and scalar multiplication on the plane.

Add or subtract vectors.

Vectors are used to measure displacement of an object or force.

Addition of vectors:
$$(a,b) + (c,d) = (a+c, b+d)$$

Addition Properties of vectors:

$$a + b = b + a$$
$$a + (b + c) = (a + b) + c$$
$$a + 0 = a$$
$$a + (^-a) = 0$$

Subtraction of vectors:

$$a - b = a + (^-b) \text{ therefore,}$$
$$a - b = (a_1, a_2) + (^-b_1, ^-b_2) \text{ or}$$
$$a - b = (a_1 - b_1, a_2 - b_2)$$

Sample problem:

If $a = (4, ^-1)$ and $b = (^-3, 6)$, find $a + b$ and $a - b$.

Using the rule for addition of vectors:
$$(4, ^-1) + (^-3, 6) = (4 + (^-3), ^-1 + 6)$$
$$= (1, 5)$$

Using the rule for subtraction of vectors:
$$(4, ^-1) - (^-3, 6) = (4 - (^-3), ^-1 - 6)$$
$$= (7, ^-7)$$

Find the dot product of two vectors.

The dot product $a \cdot b$:

$$a = (a_1, a_2) = a_1 i + a_2 j \quad \text{and} \quad b = (b_1, b_2) = b_1 i + b_2 j$$

$$a \cdot b = a_1 b_1 + a_2 b_2$$

$a \cdot b$ is read "a dot b". Dot products are also called scalar or inner products. When discussing dot products, it is important to remember that "a dot b" is not a vector, but a real number.

Properties of the dot product:

$a \cdot a = |a|^2$
$a \cdot b = b \cdot a$
$a \cdot (b + c) = a \cdot b + a \cdot c$
$(ca) \cdot b = c(a \cdot b) = a \cdot (cb)$
$0 \cdot a = 0$

Sample problems:

Find the dot product.

1. $a = (5, 2), b = (^-3, 6)$
   $a \cdot b = (5)(^-3) + (2)(6)$
   $\quad = {}^-15 + 12$
   $\quad = {}^-3$

2. $a = (5i + 3j), b = (4i - 5j)$
   $a \cdot b = (5)(4) + (3)(^-5)$
   $\quad = 20 - 15$
   $\quad = 5$

3. The magnitude and direction of a constant force are given by $a = 4i + 5j$. Find the amount of work done if the point of application of the force moves from the origin to the point $P(7, 2)$.

The work W done by a constant force a as its point of application moves along a vector b is $W = a \cdot b$.

Sketch the constant force vector a and the vector b.

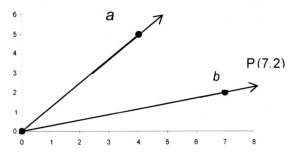

$b = (7,2) = 7i + 2j$

Use the definition of work done to solve.

$$\begin{aligned} W &= a \cdot b \\ &= (4i + 5j)(7i + 2j) \\ &= (4)(7) + (5)(2) \\ &= (28) + (10) \\ &= 38 \end{aligned}$$

SKILL 1.25 Solve real-world problems involving ratio or proportion.

A **ratio** is a comparison of 2 numbers. If a class had 11 boys and 14 girls, the ratio of boys to girls could be written one of 3 ways:

$$11:14 \quad \text{or} \quad 11 \text{ to } 14 \quad \text{or} \quad \frac{11}{14}$$

The ratio of girls to boys is:

$$14:11, \; 14 \text{ to } 11 \text{ or } \frac{14}{11}$$

Ratios can be reduced when possible. A ratio of 12 cats to 18 dogs would reduce to 2:3, 2 to 3 or $2/3$.

Note: Read ratio questions carefully. Given a group of 6 adults and 5 children, the ratio of children to the entire group would be 5:11.

A **proportion** is an equation in which a fraction is set equal to another. To solve the proportion, multiply each numerator times the other fraction's denominator. Set these two products equal to each other and solve the resulting equation. This is called **cross-multiplying** the proportion.

Example: $\dfrac{4}{15} = \dfrac{x}{60}$ is a proportion.

To solve this, cross multiply.

$$(4)(60) = (15)(x)$$
$$240 = 15x$$
$$16 = x$$

Example: $\dfrac{x+3}{3x+4} = \dfrac{2}{5}$ is a proportion.

To solve, cross multiply.

$$5(x+3) = 2(3x+4)$$
$$5x + 15 = 6x + 8$$
$$7 = x$$

Example: $\dfrac{x+2}{8} = \dfrac{2}{x-4}$ is another proportion.

To solve, cross multiply.

$$(x+2)(x-4) = 8(2)$$
$$x^2 - 2x - 8 = 16$$
$$x^2 - 2x - 24 = 0$$
$$(x-6)(x+4) = 0$$
$$x = 6 \text{ or } x = {}^-4$$

Both answers work.

TEACHER CERTIFICATION STUDY GUIDE

COMPETENCY 2.0 KNOWLEDGE OF FUNCTIONS

SKILL 2.1 Interpret the language and notation of functions.

A function can be defined as a set of ordered pairs in which each element of the domain is paired with one and only one element of the range. The symbol $f(x)$ is read "f of x." A Letter other than "f" can be used to represent a function. The letter "g" is commonly used as in $g(x)$.

Sample problems:

1. Given $f(x) = 4x^2 - 2x + 3$, find $f(^-3)$.
(This question is asking for the range value that corresponds to the domain value of $^-3$).

$$f(x) = 4x^2 - 2x + 3$$
$$f(^-3) = 4(^-3)^2 - 2(^-3) + 3$$
$$f(^-3) = 45$$

1. Replace x with $^-3$.
2. Solve.

2. Find $f(3)$ and $f(10)$, given $f(x) = 7$.

$$f(x) = 7$$
$$(3) = 7$$

1. There are no x values to substitute for. This is your answer.

$$f(x) = 7$$
$$f10) = 7$$

2. Same as above.

Notice that both answers are equal to the constant given.

SKILL 2.2 Determine which relations are functions, given mappings, sets of ordered pairs, rules, and graphs.

- A **relation** is any set of ordered pairs.

- The **domain** of a relation is the set made of all the first coordinates of the ordered pairs.

- The **range** of a relation is the set made of all the second coordinates of the ordered pairs.

- A **function** is a relation in which different ordered pairs have different first coordinates. (No x values are repeated.)

- A **mapping** is a diagram with arrows drawn from each element of the domain to the corresponding elements of the range. If 2 arrows are drawn from the same element of the domain, then it is not a function.

- On a graph, use the **vertical line test** to look for a function. If any vertical line intersects the graph of a relation in more than one point, then the relation is not a function.

1. Determine the domain and range of this mapping.

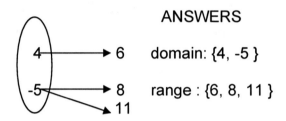

ANSWERS

domain: {4, -5 }

range : {6, 8, 11 }

2. Determine which of these are functions:

a. $\{(1,^-4),(27,1)(94,5)(2,^-4)\}$

b. $f(x) = 2x - 3$

c. $A = \{(x,y) \mid xy = 24\}$

d. $y = 3$

e. $x = {}^-9$

f. $\{(3,2),(7,7),(0,5),(2,^-4),(8,^-6),(1,0),(5,9),(6,^-4)\}$

3. Determine the domain and range of this graph.

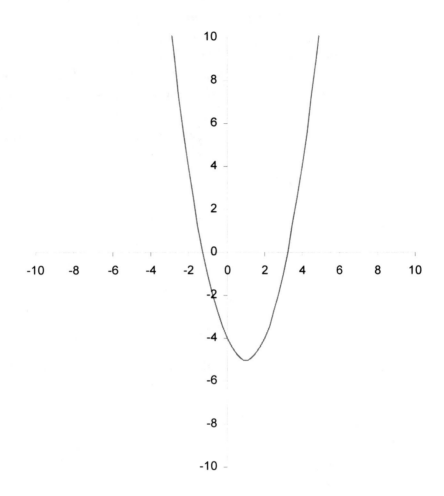

SKILL 2.3 Identify the domain and range of a given function.

- A **relation** is any set of ordered pairs.
- The **domain** of the relation is the set of all first co-ordinates of the ordered pairs. (These are the x coordinates.)
- The **range** of the relation is the set of all second co-ordinates of the ordered pairs. (These are the y coordinates.)

1. If $A = \{(x,y) \mid y = x^2 - 6\}$, find the domain and range.

2. Give the domain and range of set B if:

   $$B = \{(1, {}^-2), (4, {}^-2), (7, {}^-2), (6, {}^-2)\}$$

3. Determine the domain of this function:

   $$f(x) = \frac{5x + 7}{x^2 - 4}$$

4. Determine the domain and range of these graphs.

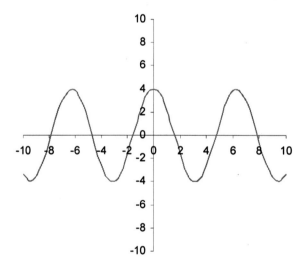

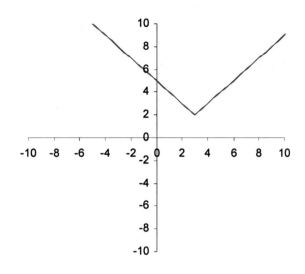

5. If $E = \{(x,y) \mid y = 5\}$, find the domain and range.

6. Determine the ordered pairs in the relation shown in this mapping.

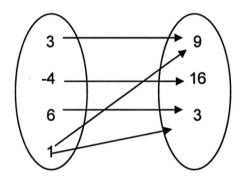

SKILL 2.4   Identify the graph of special functions (i.e., absolute value, step, piecewise, identity, constant function).

-The **absolute value function** for a 1st degree equation is of the form:
$y = m(x-h)+k$. Its graph is in the shape of a $\vee$. The point (h,k) is the location of the maximum/minimum point on the graph. "$\pm m$" are the slopes of the 2 sides of the $\vee$. The graph opens up if m is positive and down if m is negative.

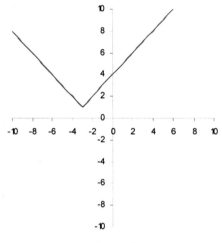

$$y = |x+3|+1$$

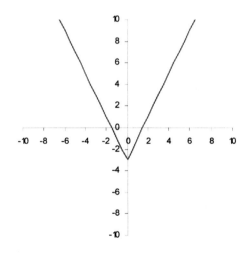

$$y = 2|x|-3$$

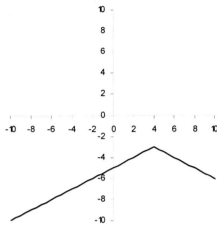

$$y = {}^-1/2|x-4|-3$$

-Note that on the first graph, the graph opens up since m is positive 1. It has ($^-$3,1) as its minimum point. The slopes of the 2 upward rays are ± 1.
-The second graph also opens up since m is positive. (0, $^-$3) is its minimum point. The slopes of the 2 upward rays are ± 2.
-The third graph is a downward $\wedge$ because m is $^-1/2$. The maximum point on the graph is at (4, $^-$3). The slopes of the 2 downward rays are $\pm 1/2$.

-The **identity function** is the linear equation $y = x$. Its graph is a line going through the origin (0,0) and through the first and third quadrants at a 45° degree angle.

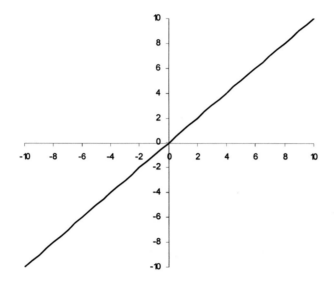

-The **greatest integer function** or **step function** has the equation: $f(x) = j[rx - h] + k$ or $y = j[rx - h] + k$. (h,k) is the location of the left endpoint of one step. j is the vertical jump from step to step. r is the reciprocal of the length of each step. If (x,y) is a point of the function, then when x is an integer, its y value is the same integer. If (x,y) is a point of the function, then when x is not an integer, its y value is the first integer less than x. Points on $y = [x]$ would include:

(3,3), (⁻2,⁻2), (0,0), (1.5,1), (2.83,2), (⁻3.2,⁻4), (⁻.4,⁻1).

$y = [x]$

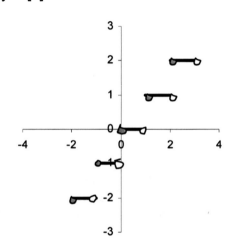

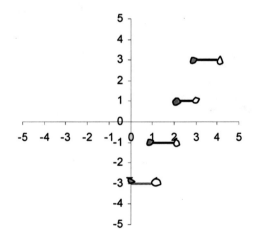

$y = 2[x] - 3$

-Note that in the graph of the first equation, the steps are going up as they move to the right. Each step is one space wide (inverse of r) with a solid dot on the left and a hollow dot on the right where the jump to the next step occurs. Each step is one square higher (j = 1) than the previous step. One step of the graph starts at $(0,0) \leftarrow$ values of (h,k).

-In the second graph, the graph goes up to the right. One step starts at the point $(0, ^-3) \leftarrow$ values of (h,k). Each step is one square wide (r = 1) and each step is 2 squares higher than the previous step (j = 2).

Practice: Graph the following equations:

1. $f(x) = x$
2. $y = ^-|x - 3| + 5$
3. $y = 3[x]$
4. $y = 2/5|x - 5| - 2$

Functions defined by two or more formulas are **piecewise functions**. The formula used to evaluate piecewise functions varies depending on the value of x. The graphs of piecewise functions consist of two or more pieces, or intervals, and are often discontinuous.

Example 1:

$f(x) = \begin{array}{ll} x + 1 & \text{if } x > 2 \\ x - 2 & \text{if } x \leq 2 \end{array}$

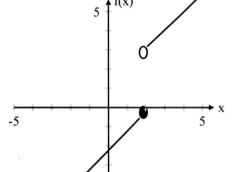

Example 2:

$f(x) = \begin{array}{ll} x & \text{if } x \geq 1 \\ x^2 & \text{if } x < 1 \end{array}$

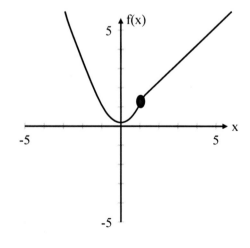

When graphing or interpreting the graph of piecewise functions it is important to note the points at the beginning and end of each interval because the graph must clearly indicate what happens at the end of each interval. Note that in the graph of Example 1, point (2, 3) is not part of the graph and is represented by an empty circle. On the other hand, point (2, 0) is part of the graph and is represented as a solid circle. Note also that the graph of Example 2 is continuous despite representing a piecewise function.

Practice: Graph the following piecewise equations.

1. $f(x) = x^2 \quad$ if $x > 0$
   $ = x + 4 \quad$ if $x \leq 0$

2. $f(x) = x^2 - 1 \quad$ if $x > 2$
   $ = x^2 + 2 \quad$ if $x \leq 2$

SKILL 2.5 Find specific values of a given function.

There are 2 easy ways to find the values of a function. First to find the value of a function when $x = 3$, substitute 3 in place of every letter x. Then simplify the expression following the order of operations. For example, if $f(x) = x^3 - 6x + 4$, then to find $f(3)$, substitute 3 for x.

The equation becomes $f(3) = 3^3 - 6(3) + 4 = 27 - 18 + 4 = 13$.
So (3, 13) is a point on the graph of $f(x)$.

A second way to find the value of a function is to use synthetic division. To find the value of a function when $x = 3$, divide 3 into the coefficients of the function. (Remember that coefficients of missing terms, like x^2, must be included). The remainder is the value of the function.

If $f(x) = x^3 - 6x + 4$, then to find $f(3)$ using synthetic division:

Note the 0 for the missing x^2 term.

$$\begin{array}{r|rrrr} 3 & 1 & 0 & -6 & 4 \\ & & 3 & 9 & 9 \\ \hline & 1 & 3 & 3 & 13 \end{array}$$ ← this is the value of the function.

Therefore, (3, 13) is a point on the graph of $f(x) = x^3 - 6x + 4$.

Example: Find values of the function at integer values from $x = -3$ to $x = 3$ if $f(x) = x^3 - 6x + 4$.

If $x = {}^-3$:

$$f({}^-3) = ({}^-3)^3 - 6({}^-3) + 4$$
$$= ({}^-27) - 6({}^-3) + 4$$
$$= {}^-27 + 18 + 4 = {}^-5$$

synthetic division:

```
     | 1   0   -6   4
  -3 |    -3    9  -9
     |_____
       1  -3    3  -5
```

$1 \; {}^-3 \; 3 \; {}^-5 \leftarrow$ this is the value of the function if $x = {}^-3$.
Therefore, $({}^-3, {}^-5)$ is a point on the graph.

If $x = {}^-2$:

$$f({}^-2) = ({}^-2)^3 - 6({}^-2) + 4$$
$$= ({}^-8) - 6({}^-2) + 4$$
$$= {}^-8 + 12 + 4 = 8 \leftarrow \text{this is the value of the function if } x = {}^-2.$$
Therefore, $({}^-2, 8)$ is a point on the graph.

If $x = {}^-1$:

$$f({}^-1) = ({}^-1)^3 - 6({}^-1) + 4$$
$$= ({}^-1) - 6({}^-1) + 4$$
$$= {}^-1 + 6 + 4 = 9$$

synthetic division:

```
     | 1   0   -6   4
  -1 |    -1    1   5
     |_____
       1  -1   -5   9
```

$1 \; {}^-1 \; {}^-5 \; 9 \leftarrow$ this is the value if the function if $x = {}^-1$.
Therefore, $({}^-1, 9)$ is a point on the graph.

If $x = 0$:
$$f(0) = (0)^3 - 6(0) + 4$$
$$= 0 - 6(0) + 4$$
$$= 0 - 0 + 4 = 4 \leftarrow \text{this is the value of the function if } x = 0.$$
Therefore, $(0, 4)$ is a point on the graph.

If $x = 1$:

$$f(1) = (1)^3 - 6(1) + 4$$
$$= (1) - 6(1) + 4$$
$$= 1 - 6 + 4 = {}^-1$$

synthetic division:

$$\begin{array}{r|rrrr} 1 & 1 & 0 & {}^-6 & 4 \\ & & 1 & 1 & {}^-5 \\ \hline & 1 & 1 & {}^-5 & {}^-1 \end{array}$$

$1 \quad 1 \quad {}^-5 \quad {}^-1 \leftarrow$ this is the value if the function of $x = 1$.
Therefore, $(1, {}^-1)$ is a point on the graph.

If $x = 2$:

$$f(2) = (2)^3 - 6(2) + 4$$
$$= 8 - 6(2) + 4$$
$$= 8 - 12 + 4 = 0$$

synthetic division:

$$\begin{array}{r|rrrr} 2 & 1 & 0 & {}^-6 & 4 \\ & & 2 & 4 & {}^-4 \\ \hline & 1 & 2 & {}^-2 & 0 \end{array}$$

$1 \quad 2 \quad {}^-2 \quad 0 \leftarrow$ this is the value of the function if $x = 2$.
Therefore, $(2, 0)$ is a point on the graph.

If $x = 3$:

$$f(3) = (3)^3 - 6(3) + 4$$
$$= 27 - 6(3) + 4$$
$$= 27 - 18 + 4 = 13$$

synthetic division:

$$\begin{array}{r|rrrr} 3 & 1 & 0 & {}^-6 & 4 \\ & & 3 & 9 & 9 \\ \hline & 1 & 3 & 3 & 13 \end{array}$$

$1 \quad 3 \quad 3 \quad 13 \leftarrow$ this is the value of the function if $x = 3$.
Therefore, $(3, 13)$ is a point on the graph.

The following points are points on the graph:

X	Y
−3	−5
−2	8
−1	9
0	4
1	−1
2	0
3	13

Note the change in sign of the y value between $x = {}^-3$ and $x = {}^-2$. This indicates there is a zero between $x = {}^-3$ and $x = {}^-2$. Since there is another change in sign of the y value between $x = 0$ and $x = {}^-1$, there is a second root there. When $x = 2$, $y = 0$ so $x = 2$ is an exact root of this polynomial.

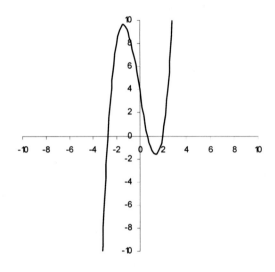

Example: Find values of the function at $x = {}^-5, 2,$ and 17 if
$f(x) = 2x^5 - 4x^3 + 3x^2 - 9x + 10$.

If $x = {}^-5$:
$f({}^-5) = 2({}^-5)^5 - 4({}^-5)^3 + 3({}^-5)^2 - 9({}^-5) + 10$
$= 2({}^-3125) - 4({}^-125) + 3(25) - 9({}^-5) + 10$
$= {}^-6250 + 500 + 75 + 45 + 10 = {}^-5620$

synthetic division:

```
   |  2   0   -4    3    -9     10
-5 |    -10  50  -230  1135  -5630
   ----------------------------------
      2  -10  46  -227  -1126  -5620  ← this is the value of the function if x = -5.
```

Therefore, $({}^-5, {}^-5620)$ is a point on the graph.

If $x = 2$:
$f(2) = 2(2)^5 - 4(2)^3 + 3(2)^2 - 9(2) + 10$
$= 2(32) - 4(8) + 3(4) - 9(2) + 10$
$= 64 - 32 + 12 - 18 + 10 = 36$

synthetic division:

```
  |  2   0   -4   3   -9   10
2 |      4    8   8   22   26
  --------------------------------
     2   4    4  11   13   36  ← this is the value of the function if x = 2.
```

Therefore, $(2, 36)$ is a point on the graph.

If $x = 17$:

$f(17) = 2(17)^5 - 4(17)^3 + 3(17)^2 - 9(17) + 10$
$= 2(1419857) - 4(4913) + 3(289) - 9(17) + 10$
$= 2839714 - 19652 + 867 - 153 + 10 = 2820786$

synthetic division:

```
    |  2   0    -4     3     -9      10
 17 |     34   578  9758  165937  2820776
    ------------------------------------------
       2  34  574  9761  165928  2820786  ← this is the value
                                            of the function if x = 17.
```

Therefore, $(17, 2820786)$ is a point on the graph.

SKILL 2.6 Estimate or find the zeros of a polynomial function.

Synthetic division can be used to find the value of a function at any value of x. To do this, divide the value of x into the coefficients of the function.

(Remember that coefficients of missing terms, like x^2 below, must be included.) The remainder of the synthetic division is the value of the function. If $f(x) = x^3 - 6x + 4$, then to find the value of the function at $x = 3$, use synthetic division:

Note the 0 for the missing x^2 term.

$$
\begin{array}{r|rrrr}
3 & 1 & 0 & -6 & 4 \\
  &   & 3 & 9 & 9 \\
\hline
  & 1 & 3 & 3 & 13
\end{array}
$$
← This is the value of the function.

Therefore, (3, 13) is a point on the graph.

<u>Example</u>: Find values of the function at $x = {}^-5$ if
$f(x) = 2x^5 - 4x^3 + 3x^2 - 9x + 10$.

Note the 0 below for the missing x^4 term.

Synthetic division:

$$
\begin{array}{r|rrrrrr}
-5 & 2 & 0 & -4 & 3 & -9 & 10 \\
   &   & -10 & 50 & -230 & 1135 & -5630 \\
\hline
   & 2 & -10 & 46 & -227 & 1126 & -5620
\end{array}
$$
← This is the value of the function if $x = {}^-5$.

Therefore, $({}^-5, {}^-5620)$ is a point on the graph.

Note that if $x = {}^-5$, the same value of the function can also be found by substituting ${}^-5$ in place of x in the function.

$$f({}^-5) = 2({}^-5)^5 - 4({}^-5)^3 + 3({}^-5)^2 - 9({}^-5) + 10$$
$$= 2({}^-3125) - 4({}^-125) + 3(25) - 9({}^-5) + 10$$
$$= {}^-6250 + 500 + 75 + 45 + 10 = {}^-5620$$

Therefore, $({}^-5, {}^-5620)$ is still a point on the graph.

To determine if $(x-a)$ or $(x+a)$ is a factor of a polynomial, do a synthetic division, dividing by the opposite of the number inside the parentheses. To see if $(x-5)$ is a factor of a polynomial, divide it by 5. If the remainder of the synthetic division is zero, then the binomial is a factor of the polynomial.

If $f(x) = x^3 - 6x + 4$, determine if $(x-1)$ is a factor of $f(x)$. Use synthetic division and divide by 1:

Note the 0 for the missing x^2 term.

$$\begin{array}{r|rrrr} 1 & 1 & 0 & ^-6 & 4 \\ & & 1 & 1 & ^-5 \\ \hline & 1 & 1 & ^-5 & ^-1 \end{array}$$ ← This is the remainder of the function.

Therefore, $(x-1)$ is **not** a factor of $f(x)$.

If $f(x) = x^3 - 6x + 4$, determine if $(x-2)$ is a factor of $f(x)$. Use synthetic division and divide by 2:

$$\begin{array}{r|rrrr} 2 & 1 & 0 & ^-6 & 4 \\ & & 2 & 4 & ^-4 \\ \hline & 1 & 2 & ^-2 & 0 \end{array}$$ ← This is the remainder of the function.

Therefore, $(x-2)$ **is** a factor of $f(x)$.

The converse of this is also true. If you divide by k in any synthetic division and get a remainder of zero for the division, then $(x-k)$ is a factor of the polynomial. Similarly, if you divide by ^-k in any synthetic division and get a remainder of zero for the division, then $(x+k)$ is a factor of the polynomial.

Divide $2x^3 - 6x - 104$ by 4. What is your conclusion?

$$\begin{array}{r|rrrr} 4 & 2 & 0 & ^-6 & ^-104 \\ & & 8 & 32 & 104 \\ \hline & 2 & 8 & 26 & 0 \end{array}$$ ← This is the remainder of the function.

Since the remainder is **0**, then **$(x-4)$** is a factor.

Given any polynomial, be sure that the exponents on the terms are in descending order. List out all of the factors of the first term's coefficient and of the constant in the last term. Make a list of fractions by putting each of the factors of the last term's coefficient over each of the factors of the first term. Reduce fractions when possible. Put a ± in front of each fraction. This list of fractions is a list of the only possible rational roots of a function. If the polynomial is of degree **n**, then at most n of these will actually be roots of the polynomial.

Example: List the possible rational roots for the function $f(x) = x^2 - 5x + 4$.

$$\pm \frac{\text{factors of 4}}{\text{factors of 1}} = \pm 1, 2, 4 \leftarrow 6 \text{ possible rational roots}$$

Example: List the possible rational roots for the function $f(x) = 6x^2 - 5x - 4$.

Make fractions of the form :

possible rational roots $= \pm \dfrac{\text{factors of 4}}{\text{factors of 6}} = \pm \dfrac{1,2,4}{1,2,3,6} =$

$\pm \dfrac{1}{2}, \dfrac{1}{3}, \dfrac{1}{6}, \dfrac{2}{3}, \dfrac{4}{3}, 1, 2, 4$

are the only 16 rational numbers that could be roots.

Since this equation is of degree 2, there are, at most, 2 rational roots. (They happen to be 4/3 and ⁻1/2.)

SKILL 2.7 Identify the sum, difference, product, and quotient of functions.

If $f(x)$ is a function and the value of 3 is in the domain, the corresponding element in the range would be f(3). It is found by evaluating the function for $x = 3$. The same holds true for adding, subtracting, and multiplying in function form.

The symbol f^{-1} is read "the inverse of f". The $^{-1}$ is not an exponent. The inverse of a function can be found by reversing the order of coordinates in each ordered pair that satisfies the function. Finding the inverse functions means switching the place of x and y and then solving for y.

Sample problem:
1. Find $p(a+1) + 3\{p(4a)\}$ if $p(x) = 2x^2 + x + 1$.

Find $p(a+1)$.

$p(a+1) = 2(a+1)^2 + (a+1) + 1$ Substitute $(a+1)$ for x.
$p(a+1) = 2a^2 + 5a + 4$ Solve.

Find $3\{p(4a)\}$.

$3\{p(4a)\} = 3[2(4a)^2 + (4a) + 1]$ Substitute $(4a)$ for x, multiply by 3.
$3\{p(4a)\} = 96a^2 + 12a + 3$ Solve.

$p(a+1) + 3\{p(4a)\} = 2a^2 + 5a + 4 + 96a^2 + 12a + 3$
Combine like terms.

$p(a+1) + 3\{p(4a)\} = 98a^2 + 17a + 7$

MATHEMATICS 6-12

SKILL 2.8 Determine the inverse of a given function.

How to write the equation of the inverse of a function.

1. To find the inverse of an equation using x and y, replace each letter with the other letter. Then solve the new equation for y, when possible. Given an equation like $y = 3x - 4$, replace each letter with the other:

$x = 3y - 4$. Now solve this for y:
$x + 4 = 3y$
$1/3\, x + 4/3 = y$ This is the inverse.

Sometimes the function is named by a letter:

$f(x) = 5x + 10$

Temporarily replace $f(x)$ with y.

$y = 5x + 10$

Now replace each letter with the other: $x = 5y + 10$
Solve for the new y: $x - 10 = 5y$
$1/5\, x - 2 = y$

The inverse of $f(x)$ is denoted as $f^{-1}(x)$, so the answer is
$f^{-1}(x) = 1/5\, X - 2$.

SKILL 2.9 Determine the composition of two functions.

Composition is a process that creates a new function by substituting an entire function into another function. The composition of two functions f(x) and g(x) is denoted by (f ∘ g)(x) or f(g(x)). The domain of the composed function, f(g(x)), is the set of all values of x in the domain of g that produce a value for g(x) that is in the domain of f. In other words, f(g(x)) is defined whenever both g(x) and f(g(x)) are defined.

Example 1:

If f(x) = x + 1 and g(x) = x^3, find the composition functions f ∘ g and g ∘ f and state their domains.

Solution:

(f ∘ g)(x) = f(g(x)) = f(x^3) = x^3 + 1
(g ∘ f)(x) = g(f(x)) = g(x + 1) = $(x + 1)^3$

The domain of both composite functions is the set of all real numbers.

Note that f(g(x)) and g(f(x)) are not the same. In general, unlike multiplication and addition, composition is not reversible. Thus, the order of composition is important.

Example 2:

If f(x) = sqrt(x) and g(x) = x +2, find the composition functions f ∘ g and g ∘ f and state their domains.

Solution:

(f ∘ g)(x) = f(g(x)) = f(x + 2) = sqrt(x + 2)
(g ∘ f)(x) = g(f(x)) = g(sqrt(x)) = sqrt(x) +2

The domain of f(g(x)) is x ≥ -2 because x + 2 must be non-negative in order to take the square root.

The domain of g(f(x)) is x ≥ 0 because x must be non-negative in order to take the square root.

Note that defining the domain of composite functions is important when square roots are involved.

TEACHER CERTIFICATION STUDY GUIDE

SKILL 2.10 Determine whether a function is symmetric, periodic, or even/odd.

Definition: A function f is even (symmetric about the y-axis) if $f(^-x) = f(x)$ and odd (symmetric about the origin) if $f(^-x) = {}^-f(x)$ for all x in the domain of f.

Sample problems:

Determine if the given function is even, odd, or neither even nor odd.

1. $f(x) = x^4 - 2x^2 + 7$
$f(^-x) = (^-x)^4 - 2(^-x)^2 + 7$
$f(^-x) = x^4 - 2x^2 + 7$

   f(x) is an even function.

   1. Find $f(^-x)$.
   2. Replace x with ^-x.
   3. Since $f(^-x) = f(x)$, f(x) is an even function.

2. $f(x) = 3x^3 + 2x$
$f(^-x) = 3(^-x)^3 + 2(^-x)$
$f(^-x) = {}^-3x^3 - 2x$

   $^-f(x) = {}^-(3x^3 + 2x)$
   $^-f(x) = {}^-3x^3 - 2x$

   f(x) is an odd function.

   1. Find $f(^-x)$.
   2. Replace x with ^-x.
   3. Since $f(x)$ is not equal to $f(^-x)$, f(x) is not an even function.
   4. Try $^-f(x)$.
   5. Since $f(^-x) = {}^-f(x)$, f(x) is an odd function.

3. $g(x) = 2x^2 - x + 4$
$g(^-x) = 2(^-x)^2 - (^-x) + 4$
$g(^-x) = 2x^2 + x + 4$

   $^-g(x) = {}^-(2x^2 - x + 4)$
   $^-g(x) = {}^-2x^2 + x - 4$

   $g(x)$ is neither even nor odd.

   1. First find $g(^-x)$.
   2. Replace x with ^-x.
   3. Since $g(x)$ does not equal $g(^-x)$, $g(x)$ is not an even function.
   4. Try $^-g(x)$.
   5. Since $^-g(x)$ does not equal $g(^-x)$, $g(x)$ is not an odd function.

MATHEMATICS 6-12

TEACHER CERTIFICATION STUDY GUIDE

SKILL 2.11 Determine the graph of the image of a function under given transformations (i.e., translation, rotations through multiples of 90 degrees, dilations, and/or reflections over y=x horizontal or vertical lines).

Different types of function transformations affect the graph and characteristics of a function in predictable ways. The basic types of transformation are horizontal and vertical shift (translation), horizontal and vertical scaling (dilation), and reflection. As an example of the types of transformations, we will consider transformations of the functions $f(x) = x^2$.

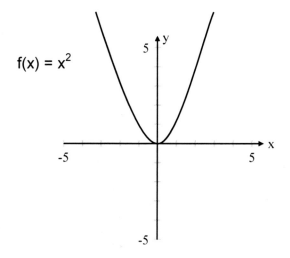

Horizontal shifts take the form $g(x) = f(x \pm c)$. For example, we obtain the graph of the function $g(x) = (x + 2)^2$ by shifting the graph of $f(x) = x^2$ two units to the left. The graph of the function $h(x) = (x - 2)^2$ is the graph of $f(x) = x^2$ shifted two units to the right.

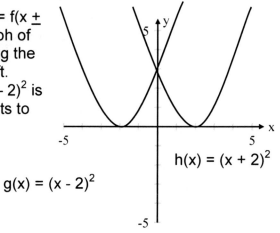

MATHEMATICS 6-12

Vertical shifts take the form $g(x) = f(x) \pm c$. For example, we obtain the graph of the function $g(x) = (x^2) - 2$ by shifting the graph of $f(x) = x^2$ two units down. The graph of the function $h(x) = (x^2) + 2$ is the graph of $f(x) = x^2$ shifted two units up.

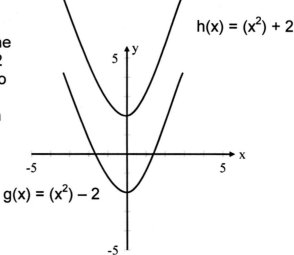

Horizontal scaling takes the form $g(x) = f(cx)$. For example, we obtain the graph of the function $g(x) = (2x)^2$ by compressing the graph of $f(x) = x^2$ in the x-direction by a factor of two. If $c > 1$ the graph is compressed in the x-direction, while if $1 > c > 0$ the graph is stretched in the x-direction.

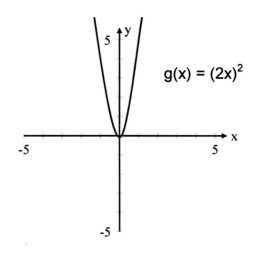

COMPETENCY 3.0 KNOWLEDGE OF GEOMETRY FROM A SYNTHETIC PERSPECTIVE

SKILL 3.1 Determine the change in the area or volume of a figure when its dimensions are altered.

Examining the change in area or volume of a given figure requires first to find the existing area given the original dimensions and then finding the new area given the increased dimensions.

Sample problem:

Given the rectangle below determine the change in area if the length is increased by 5 and the width is increased by 7.

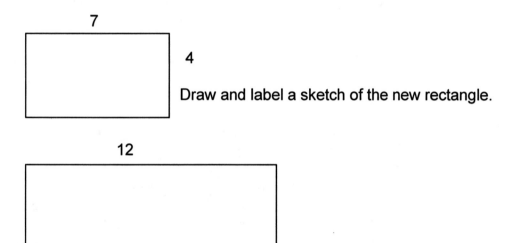

Draw and label a sketch of the new rectangle.

Find the areas.

Area of original = LW
    = (7)(4)
    = 28 units^2

Area of enlarged shape = LW
    = (12)(11)
    = 132 units^2

The change in area is 132 − 28 = 104 units^2.

SKILL 3.2 Estimate measurements of familiar objects using metric or standard units.

It is necessary to be familiar with the metric and customary system in order to estimate measurements.

Some common equivalents include:

ITEM	APPROXIMATELY EQUAL TO:	
	METRIC	IMPERIAL
large paper clip	1 gram	1 ounce
1 quart	1 liter	
average sized man	75 kilograms	170 pounds
1 yard	1 meter	
math textbook	1 kilogram	2 pounds
1 mile	1 kilometer	
1 foot	30 centimeters	
thickness of a dime	1 millimeter	0.1 inches

Estimate the measurement of the following items:

The length of an adult cow = _____ meters
The thickness of a compact disc = _____ millimeters
Your height = _____ meters
length of your nose = _____ centimeters
weight of your math textbook = _____ kilograms
weight of an automobile = _____ kilograms
weight of an aspirin = _____ grams

SKILL 3.3 Determine the relationships between points, lines, and planes, including their intersections.

In geometry the point, line and plane are key concepts and can be discussed in relation to each other.

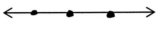

collinear points
are all on the same line

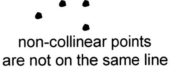

non-collinear points
are not on the same line

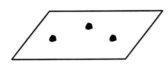

coplanar points
are on the same plane

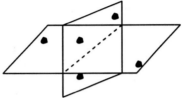

non-coplanar points
are not on the same plane

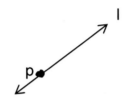

Point p is in line l
Point p is on line l
l contains P
l passes through P

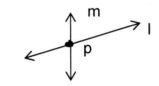

l and m intersect
at p
p is the intersection
of l and m

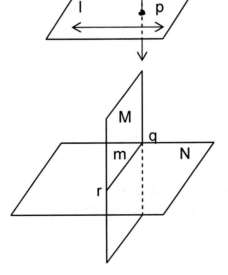

l and p are in plane N
N contains p and l
m intersects N at p
p is the intersection
of m and N

Planes M and N intersect at rq
rq is the intersection
of M and N
rq is in M and N
M and N contain rQ

SKILL 3.4 Classify geometric figures (e.g., lines, planes, angles, polygons, solids) according to their properties.

Lines and planes

Parallel lines or planes do not intersect.

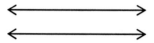

Perpendicular lines or planes form a 90 degree angle to each other.

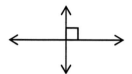

Intersecting lines share a common point and intersecting planes share a common set of points or line.

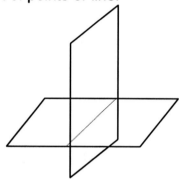

Skew lines do not intersect and do not lie on the same plane.

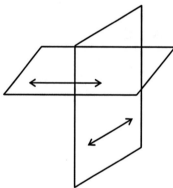

Angles

The classifying of angles refers to the angle measure. The naming of angles refers to the letters or numbers used to label the angle.

Sample Problem:

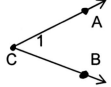

$\overrightarrow{CA}$ (read ray CA) and $\overrightarrow{CB}$ are the sides of the angle.
The angle can be called $\angle ACB$, $\angle BCA$, $\angle C$ or $\angle 1$.

Angles are classified according to their size as follows:

acute: greater than 0 and less than 90 degrees.
right: exactly 90 degrees.
obtuse: greater than 90 and less than 180 degrees.
straight: exactly 180 degrees

Angle relationships

Angles can be classified in a number of ways. Some of those classifications are outlined here.

Adjacent angles have a common vertex and one common side but no interior points in common.

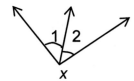

Complimentary angles add up to 90 degrees.

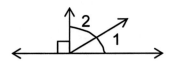

Supplementary angles add up to 180 degrees.

Vertical angles have sides that form two pairs of opposite rays.

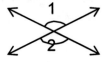

Corresponding angles are in the same corresponding position on two parallel lines cut by a transversal.

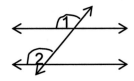

Alternate interior angles are diagonal angles on the inside of two parallel lines cut by a transversal.

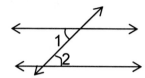

Alternate exterior angles are diagonal on the outside of two parallel lines cut by a transversal.

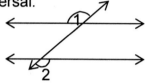

Triangles

A **triangle** is a polygon with three sides.

Triangles can be classified by the types of angles or the lengths of their sides.

Classifying by angles:

An **acute** triangle has exactly three *acute* angles.
A **right** triangle has one *right* angle.
An **obtuse** triangle has one *obtuse* angle.

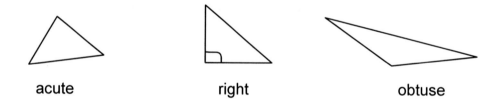

acuterightobtuse

Classifying by sides:

All *three* sides of an **equilateral** triangle are the same length.
Two sides of an **isosceles** triangle are the same length.
None of the sides of a **scalene** triangle are the same length.

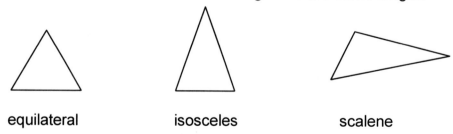

equilateralisoscelesscalene

Convex and regular polygons

In order to determine if a figure is convex and then determine if it is regular, it is necessary to apply the definition of convex first.

Convex polygons: polygons in which no line containing the side of the polygon contains a point on the interior of the polygon.

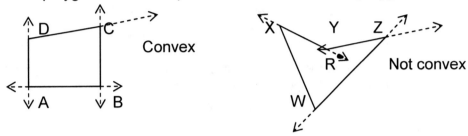

Regular polygons: convex polygons in which all sides are congruent and all angles are congruent (in other words, a regular polygon must be both equilateral and equiangular).

SKILL 3.5 Determine the measures of interior and exterior angles of any polygon.

A **polygon** is a simple closed figure composed of line segments.

In a **regular polygon** all sides are the same length and all angles are the same measure.

The sum of the measures of the **interior angles** of a polygon can be determined using the following formula, where n represents the number of angles in the polygon.

$$\text{Sum of } \angle s = 180(n - 2)$$

The measure of each angle of a regular polygon can be found by dividing the sum of the measures by the number of angles.

$$\text{Measure of } \angle = \frac{180(n-2)}{n}$$

MATHEMATICS 6-12   89

Example: Find the measure of each angle of a regular octagon.

Since an octagon has eight sides, each angle equals:

$$\frac{180(8-2)}{8} = \frac{180(6)}{8} = 135°$$

The sum of the measures of the **exterior angles** of a polygon, taken one angle at each vertex, equals 360°.

The measure of each exterior angle of a regular polygon can be determined using the following formula, where n represents the number of angles in the polygon.

Measure of exterior ∠ of regular polygon = $180 - \frac{180(n-2)}{n}$

or, more simply = $\frac{360}{n}$

Example: Find the measure of the interior and exterior angles of a regular pentagon.

Since a pentagon has five sides, each exterior angle measures:

$$\frac{360}{5} = 72°$$

Since each exterior angles is supplementary to its interior angle, the interior angle measures 180 - 72 or 108°.

SKILL 3.6 Determine the sum of the measures of the interior angles and the sum of the measures of the exterior angles of convex polygons.

A **quadrilateral** is a polygon with four sides.
The sum of the measures of the angles of a quadrilateral is 360°.

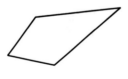

A **trapezoid** is a quadrilateral with exactly <u>one</u> pair of parallel sides.

In an **isosceles trapezoid**, the non-parallel sides are congruent.

A **parallelogram** is a quadrilateral with <u>two</u> pairs of parallel sides.

A **rectangle** is a parallelogram with a right angle.

A **rhombus** is a parallelogram with all sides equal length.

A **square** is a rectangle with all sides equal length.

SKILL 3.7 **Identify applications of special properties of trapezoids, parallelograms, and kites.**

Trapezoids

A **trapezoid** is a quadrilateral with exactly <u>one</u> pair of parallel sides.

The median of a trapezoid is parallel to the two bases.

The length of the median is equal to one-half the sum of the length of the two bases.

In an **isosceles trapezoid**, the non-parallel sides are congruent.

An isosceles trapezoid has the following properties:

The diagonals of an isosceles trapezoid are congruent.
The base angles of an isosceles trapezoid are congruent.

Example:

An isosceles trapezoid has a diagonal of 10 and a base angle measure of 30°. Find the measure of the other 3 angles.

Based on the properties of trapezoids, the measure of the other base angle is 30° and the measure of the other diagonal is 10. The other two angles have a measure of:

$$360 = 30(2) + 2x$$
$$x = 150°$$

The other two angles measure 150° each.

Parallelograms

A **parallelogram** exhibits these properties.

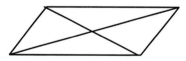

The diagonals bisect each other.
Each diagonal divides the parallelogram into two congruent triangles.
Both pairs of opposite sides are congruent.
Both pairs of opposite angles are congruent.
Two adjacent angles are supplementary.

Example 1:
Find the measures of the other three angles of a parallelogram if one angle measures 38°.

Since opposite angles are equal, there are two angles measuring 38°.

Since adjacent angles are supplementary, 180 - 38 = 142 so the other two angles measure 142° each.

```
    38
    38
   142
 + 142
   360
```

Example 2:
The measures of two adjacent angles of a parallelogram are $3x + 40$ and $x + 70$.

Find the measures of each angle.

$$2(3x + 40) + 2(x + 70) = 360$$
$$6x + 80 + 2x + 140 = 360$$
$$8x + 220 = 360$$
$$8x = 140$$
$$x = 17.5$$
$$3x + 40 = 92.5$$
$$x + 70 = 87.5$$

Thus the angles measure 92.5°, 92.5°, 87.5°, and 87.5°.

SKILL 3.8   Solve problems using the definition of congruent polygons and related theorems.

Congruent figures have the same size and shape. If one is placed above the other, it will fit exactly. Congruent lines have the same length. Congruent angles have equal measures.
The symbol for congruent is ≅.

Polygons (pentagons) ABCDE and VWXYZ are congruent. They are exactly the same size and shape.

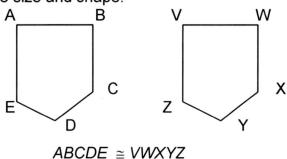

ABCDE ≅ VWXYZ

Corresponding parts are those congruent angles and congruent sides, that is:

corresponding angles	corresponding sides
$\angle A \leftrightarrow \angle V$	$AB \leftrightarrow VW$
$\angle B \leftrightarrow \angle W$	$BC \leftrightarrow WX$
$\angle C \leftrightarrow \angle X$	$CD \leftrightarrow XY$
$\angle D \leftrightarrow \angle Y$	$DE \leftrightarrow YZ$
$\angle E \leftrightarrow \angle Z$	$AE \leftrightarrow VZ$

Use the SAS, ASA, and SSS postulates to show pairs of triangles congruent.

Two triangles can be proven congruent by comparing pairs of appropriate congruent corresponding parts.

SSS POSTULATE

If three sides of one triangle are congruent to three sides of another triangle, then the two triangles are congruent.

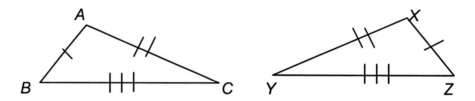

Since $AB \cong XY$, $BC \cong YZ$ and $AC \cong XZ$, then $\triangle ABC \cong \triangle XYZ$.

Example: Given isosceles triangle ABC with D the midpoint of base AC, prove the two triangles formed by AD are congruent.

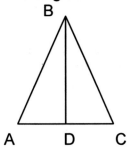

Proof:
1. Isosceles triangle ABC, D midpoint of base AC   Given
2. AB ≅ BC   An isosceles △ has two congruent sides
3. AD ≅ DC   Midpoint divides a line into two equal parts
4. BD ≅ BD   Reflexive
5. △ ABD ≅ △BCD   SSS

SAS POSTULATE

If two sides and the included angle of one triangle are congruent to two sides and the included angle of another triangle, then the two triangles are congruent.

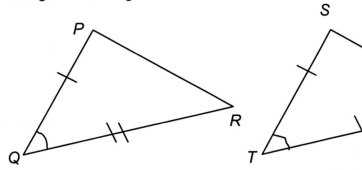

Example:

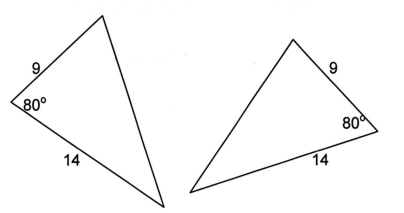

The two triangles are congruent by SAS.

ASA POSTULATE

If two angles and the included side of one triangle are congruent to two angles and the included side of another triangle, the triangles are congruent.

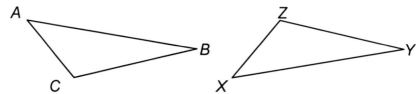

$\angle A \cong \angle X$, $\angle B \cong \angle Y$, $AB \cong XY$ then $\triangle ABC \cong \triangle XYZ$ by ASA

Example 1: Given two right triangles with one leg of each measuring 6 cm and the adjacent angle 37°, prove the triangles are congruent.

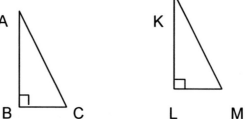

1. Right triangles ABC and KLM AB = KL = 6 cm ∠A = ∠K = 37°	Given
2. AB ≅ KL ∠A ≅ ∠K	Figures with the same measure are congruent
3. ∠B ≅ ∠L	All right angles are congruent.
4. △ABC ≅ △KLM	ASA

Example 2:
What method would you use to prove the triangles congruent?

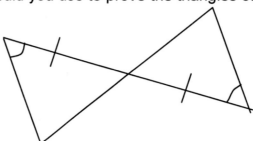

ASA because vertical angles are congruent.

AAS THEOREM

If two angles and a non-included side of one triangle are congruent to the corresponding parts of another triangle, then the triangles are congruent.

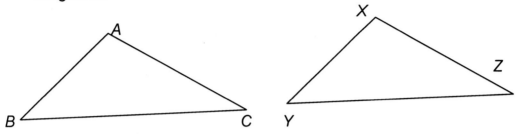

$\angle B \cong \angle Y$, $\angle C \cong \angle Z$, $AC \cong XZ$, then $\triangle ABC \cong \triangle XYZ$ by AAS.
We can derive this theorem because if two angles of the triangles are congruent, then the third angle must also be congruent. Therefore, we can use the ASA postulate.

HL THEOREM

If the hypotenuse and a leg of one right triangle are congruent to the corresponding parts of another right triangle, the triangles are congruent.

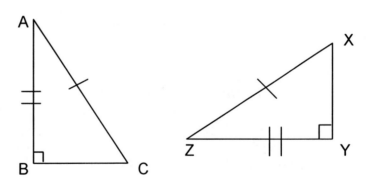

Since $\angle B$ and $\angle Y$ are right angles and $AC \cong XZ$ (hypotenuse of each triangle), $AB \cong YZ$ (corresponding leg of each triangle), then $\triangle ABC \cong \triangle XYZ$ by HL.

Example: What method would you use to prove the triangles congruent?

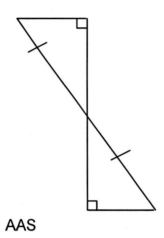

AAS

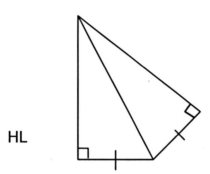

HL

SKILL 3.9 Solve problems using the definition of similar polygons and solids and related theorems.

Two figures that have the **same shape** are **similar**. Two polygons are similar if corresponding angles are congruent and corresponding sides are in proportion. Corresponding parts of similar polygons are proportional.

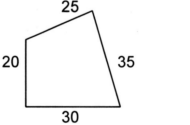 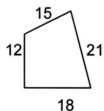

Similar Triangles

AA Similarity Postulate

If two angles of one triangle are congruent to two angles of another triangle, then the triangles are similar.

SAS Similarity Theorem

If an angle of one triangle is congruent to an angle of another triangle and the sides adjacent to those angles are in proportion, then the triangles are similar.

SSS Similarity Theorem

If the sides of two triangles are in proportion, then the triangles are similar.

Use ratios and proportions to solve problems.

Explanation can be found in Skill 26.1.

Example:

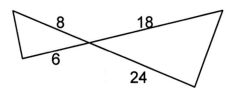

The two triangles are similar since the sides are proportional and vertical angles are congruent.

Example: Given two similar quadrilaterals. Find the lengths of sides x, y, and z.

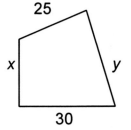 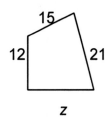

Since corresponding sides are proportional:

= so the scale is

$$\frac{12}{x} = \frac{3}{5} \qquad \frac{21}{y} = \frac{3}{5} \qquad \frac{z}{30} = \frac{3}{5}$$

$$3x = 60 \qquad 3y = 105 \qquad 5z = 90$$
$$x = 20 \qquad y = 35 \qquad z = 18$$

Similar solids share the same shape but are not necessarily the same size. The ratio of any two corresponding measurements of similar solids is the scale factor. For example, the scale factor for two square pyramids, one with a side measuring 2 inches and the other with a side measuring 4 inches, is 2:4.

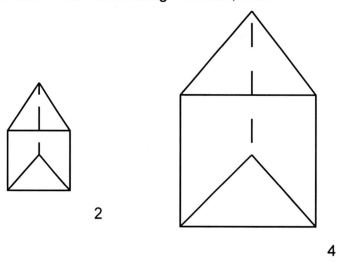

The base perimeter, the surface area, and the volume of similar solids are directly related to the scale factor. If the scale factor of two similar solids is a:b, then the...

> ratio of base perimeters = a:b
> ratio of areas = $a^2:b^2$
> ratio of volumes = $a^3:b^3$

Thus, for the above example the...

> ratio of base perimeters = 2:4
> ratio of areas = $2^2:4^2$ = 4:16
> ratio of volumes = $2^3:4^3$ = 8:64

Sample problems:

1. What happens to the volume of a square pyramid when the length of the sides of the base are doubled?

   scale factor = a:b = 1:2
   ratio of volume = $1^3:2^3$ = 1:8 (The volume is increased 8 times.)

2. Given the following measurements for two similar cylinders with a scale factor of 2:5 (Cylinders A to Cylinder B), determine the height, radius, and volume of each cylinder.

   Cylinder A: r = 2
   Cylinder B: h = 10

   Solution:

   Cylinder A –

   $$\frac{h_a}{10} = \frac{2}{5}$$
   $5h_a = 20$ Solve for h_a
   $h_a = 4$

   Volume of Cylinder a = $\pi r^2 h = \pi(2)^2 4 = 16\pi$

   Cylinder B –

   $$\frac{2}{r_b} = \frac{2}{5}$$
   $2r_b = 10$ Solve for r_b
   $r_b = 5$

   Volume of Cylinder b = $\pi r^2 h = \pi(5)^2 10 = 250\pi$

SKILL 3.10 Apply the Pythagorean theorem or its converse.

Pythagorean theorem states that the square of the length of the hypotenuse is equal to the sum of the squares of the lengths of the legs. Symbolically, this is stated as:

$$c^2 = a^2 + b^2$$

Given the right triangle below, find the missing side.

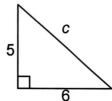

$c^2 = a^2 + b^2$ 1. write formula
$c^2 = 5^2 + 6^2$ 2. substitute known values
$c^2 = 61$ 3. take square root
$c = \sqrt{61}$ or 7.81   4. solve

Converse of the Pythagorean theorem

The Converse of the Pythagorean Theorem states that if the square of one side of a triangle is equal to the sum of the squares of the other two sides, then the triangle is a right triangle.

Example:
Given $\triangle XYZ$, with sides measuring 12, 16 and 20 cm. Is this a right triangle?

$c^2 = a^2 + b^2$
$20^2 \ ?\ 12^2 + 16^2$
$400 \ ?\ 144 + 256$
$400 = 400$

Yes, the triangle is a right triangle.

This theorem can be expanded to determine if triangles are obtuse or acute.

If the square of the longest side of a triangle is greater than the sum of the squares of the other two sides, then the triangle is an obtuse triangle.
and
If the square of the longest side of a triangle is less than the sum of the squares of the other two sides, then the triangle is an acute triangle.

Example:
Given ΔLMN with sides measuring 7, 12, and 14 inches. Is the triangle right, acute, or obtuse?

$$14^2 \; ? \; 7^2 + 12^2$$
$$196 \; ? \; 49 + 144$$
$$196 > 193$$

Therefore, the triangle is obtuse.

SKILL 3.11 **Use 30-60-90 or 45-45-90 triangle relationships to determine the lengths of the sides of triangles.**

Given the special right triangles below, we can find the lengths of other special right triangles.

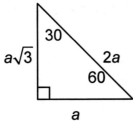

Sample problems:

1. if $8 = a\sqrt{2}$ then $a = 8/\sqrt{2}$ or 5.657

2. if $7 = a$ then $c = a\sqrt{2} = 7\sqrt{2}$ or 9.899

3. if $2a = 10$ then $a = 5$ and $x = a\sqrt{3} = 5\sqrt{3}$ or 8.66

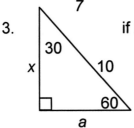

SKILL 3.12 Calculate the perimeter, circumference, and/or area of two-dimensional figures (e.g., circles, sectors, polygons, composite figures).

Circles

Given a circular figure the formulas are as follows:
$$A = \pi r^2 \qquad C = \pi d \text{ or } 2\pi r$$

Sample problem:

1. If the area of a circle is 50 cm², find the circumference.

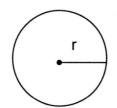

$A = 50$ cm²

1. Draw sketch.
2. Determine what is still needed.

Use the area formula to find the radius.

$A = \pi r^2$	1. write formula
$50 = \pi r^2$	2. substitute
$\dfrac{50}{\pi} = r^2$	3. divide by π
$15.915 = r^2$	4. substitute
$\sqrt{15.915} = \sqrt{r^2}$	5. take square root of both sides
$3.989 \approx r$	6. compute

Use the approximate answer (due to rounding) to find the circumference.

$C = 2\pi r$	1. write formula
$C = 2\pi (3.989)$	2. substitute
$C \approx 25.064$	3. compute

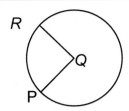

$$\frac{\angle PQR}{360°} = \frac{\text{length of arc } RP}{\text{circumference of } \odot Q} = \frac{\text{area of sector } PQR}{\text{area of } \odot Q}$$

While an arc has a measure associated to the degree measure of a central angle, it also has a length which is a fraction of the circumference of the circle.

For each central angle and its associated arc, there is a sector of the circle which resembles a pie piece. The area of such a sector is a fraction of the area of the circle.

The fractions used for the area of a sector and length of its associated arc are both equal to the ratio of the central angle to 360°.

Examples:

1.

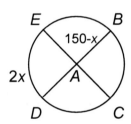

$\odot A$ has a radius of 4 cm. What is the length of arc ED?

$$2x + 150 - x = 180$$
$$x + 150 = 180$$
$$x = 30$$

Arc BE and arc DE make a semicircle.

Arc $ED = 2(30) = 60°$

The ratio 60° to 360° is equal to the ratio of arch length ED to the circumference of $\odot A$.

$$\frac{60}{360} = \frac{\text{arc length } ED}{2\pi 4}$$
$$\frac{1}{6} = \frac{\text{arc length}}{8\pi}$$

Cross multiply and solve for the arc length.

$$\frac{8\pi}{6} = \text{arc length}$$

arc length $ED = \frac{4\pi}{3}$ cm.

2.

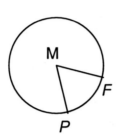

The radius of ⊙M is 3 cm. The length of arc PF is 2π cm.

What is the area of sector PMF?

Circumference of ⊙$M = 2\pi(3) = 6\pi$
Area of ⊙$M = \pi(3)^2 = 9\pi$

Find the circumference and area of the circle.

$$\frac{\text{area of } PMF}{9\pi} = \frac{2\pi}{6\pi}$$

$$\frac{\text{area of } PMF}{9\pi} = \frac{1}{3}$$

The ratio of the sector area to the circle area is the same as the arc length to the circumference.

$$\text{area of } PMF = \frac{9\pi}{3}$$

area of $PMF = 3\pi$ Solve for the area of the sector.

Polygons

FIGURE	AREA FORMULA	PERIMETER FORMULA
Rectangle	LW	$2(L+W)$
Triangle	$\frac{1}{2}bh$	$a+b+c$
Parallelogram	bh	sum of lengths of sides
Trapezoid	$\frac{1}{2}h(a+b)$	sum of lengths of sides.

Sample problems:

1. Find the area and perimeter of a rectangle if its length is 12 inches and its diagonal is 15 inches.

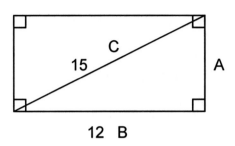

1. Draw and label sketch.
2. Since the height is still needed use Pythagorean formula to find missing leg of the triangle.

$$A^2 + B^2 = C^2$$
$$A^2 + 12^2 = 15^2$$
$$A^2 = 15^2 - 12^2$$
$$A^2 = 81$$
$$A = 9$$

Now use this information to find the area and perimeter.

$A = LW$	$P = 2(L+W)$	1. write formula
$A = (12)(9)$	$P = 2(12+9)$	2. substitute
$A = 108 \text{ in}^2$	$P = 42$ inches	3. solve

Regular polygons

Given the figure below, find the area by dividing the polygon into smaller shapes.

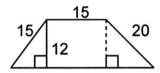

1. divide the figure into two triangles and a rectangle.
2. find the missing lengths.
3. find the area of each part.
4. find the sum of all areas.

Find base of both right triangles using Pythagorean Formula:

$$a^2 + b^2 = c^2 \qquad\qquad a^2 + b^2 = c^2$$
$$a^2 + 12^2 = 15^2 \qquad\qquad a^2 + 12^2 = 20^2$$
$$a^2 = 225 - 144 \qquad\qquad a^2 = 400 - 144$$
$$a^2 = 81 \qquad\qquad\qquad a^2 = 256$$
$$a = 9 \qquad\qquad\qquad\quad a = 16$$

Area of triangle 1	Area of triangle 2	Area of rectangle
$A = \frac{1}{2}bh$	$A = \frac{1}{2}bh$	$A = LW$
$A = \frac{1}{2}(9)(12)$	$A = \frac{1}{2}(16)(12)$	$A = (15)(12)$
$A = 54$ sq. units	$A = 96$ sq. units	$A = 180$ sq. units

Find the sum of all three figures.
$54 + 96 + 180 = 330$ square units

Parallelograms, triangles, and trapezoids

When using formulas to find each of the required items it is helpful to remember to always use the same strategies for problem solving. First, draw and label a sketch if needed. Second, write the formula down and then substitute in the known values. This will assist in identifying what is still needed (the unknown). Finally, solve the resulting equation.

Being consistent in the strategic approach to problem solving is paramount to teaching the concept as well as solving it.

Composite figures

Cut the compound shape into smaller, more familiar shapes and then compute the total area by adding the areas of the smaller parts.
Sample problem:
Find the area of the given shape.

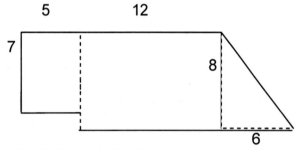

1. Using a dotted line we have cut the shape into smaller parts that are familiar.

2. Use the appropriate formula for each shape and find the sum of all areas.

Area 1 = LW   Area 2 = LW   Area 3 = ½bh
 = (5)(7)   = (12)(8)   = ½(6)(8)
 = 35 units^2 = 96 units^2 = 24 units^2

Total area = Area 1 + Area 2 + Area 3

= 35 + 96 + 24
= 155 units^2

Composite figures composed of parallelograms, triangles and trapezoids.

Use appropriate problem solving strategies to find the solution.

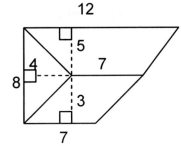

1. Find the area of the given figure.
2. Cut the figure into familiar shapes.
3. Identify what type figures are given and write the appropriate formulas.

Area of figure 1   Area of figure 2   Area of figure 3
(triangle)   (parallelogram)   (trapezoid)

$A = \frac{1}{2}bh$ $A = bh$ $A = \frac{1}{2}h(a+b)$

$A = \frac{1}{2}(8)(4)$ $A = (7)(3)$ $A = \frac{1}{2}(5)(12+7)$

$A = 16$ sq. ft $A = 21$ sq. ft $A = 47.5$ sq. ft

Now find the total area by adding the area of all figures.
Total area = 16 + 21 + 47.5
Total area = 84.5 square ft

SKILL 3.13 Apply the theorems pertaining to the relationships of chords, secants, diameters, radii, and tangents with respect to circles and to each other.

A tangent line intersects a circle in exactly one point. If a radius is drawn to that point, the radius will be perpendicular to the tangent.

A chord is a segment with endpoints on the circle. If a radius or diameter is perpendicular to a chord, the radius will cut the chord into two equal parts.

If two chords in the same circle have the same length, the two chords will have arcs that are the same length, and the two chords will be equidistant from the center of the circle. Distance from the center to a chord is measured by finding the length of a segment from the center perpendicular to the chord.

Examples:

1.

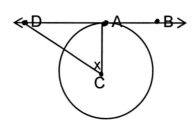

$\overrightarrow{DB}$ is tangent to $\odot C$ at A.

$m\angle ADC = 40°$. Find x.

$\overline{AC} \perp \overrightarrow{DB}$ A radius is $\perp$ to a tangent at the point of tangency.

$m\angle DAC = 90°$ Two segments that are $\perp$ form a $90°$ angle.

$40 + 90 + x = 180$ The sum of the angles of a triangle is $180°$.

$x = 50°$ Solve for x.

2.

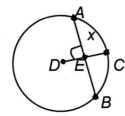

$\overline{CD}$ is a radius and $\overline{CD} \perp$ chord $\overline{AB}$.
$\overline{AB} = 10$. Find x.

$x = \frac{1}{2}(10)$

$x = 5$ If a radius is $\perp$ to a chord, the radius bisects the chord.

Lengths of chords, secants, and tangents

Intersecting chords:

If two chords intersect inside a circle, each chord is divided into two smaller segments. The product of the lengths of the two segments formed from one chord equals the product of the lengths of the two segments formed from the other chord.

Intersecting tangent segments:

If two tangent segments intersect outside of a circle, the two segments have the same length.

Intersecting secant segments:

If two secant segments intersect outside a circle, a portion of each segment will lie inside the circle and a portion (called the exterior segment) will lie outside the circle. The product of the length of one secant segment and the length of its exterior segment equals the product of the length of the other secant segment and the length of its exterior segment.

Tangent segments intersecting secant segments:

If a tangent segment and a secant segment intersect outside a circle, the square of the length of the tangent segment equals the product of the length of the secant segment and its exterior segment.

Examples:

1.
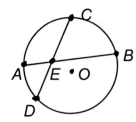

$\overline{AB}$ and $\overline{CD}$ are chords.
CE=10, ED=x, AE=5, EB=4

$(AE)(EB) = (CE)(ED)$ Since the chords intersect in the circle,
$5(4) = 10x$ the products of the segment pieces are
$20 = 10x$ equal.
$x = 2$ Solve for x.

2.
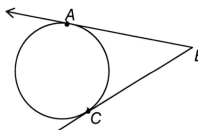

$\overline{AB}$ and $\overline{CD}$ are chords.
$\overline{AB} = x^2 + x - 2$
$\overline{BC} = x^2 - 3x + 5$

Find the length of $\overline{AB}$ and $\overline{BC}$.

$\overline{AB} = x^2 + x - 2$
$\overline{BC} = x^2 - 3x + 5$ Given

$\overline{AB} = \overline{BC}$ Intersecting tangents are equal.

$x^2 + x - 2 = x^2 - 3x + 5$ Set the expression equal and solve.

$4x = 7$
$x = 1.75$ Substitute and solve.

$(1.75)^2 + 1.75 - 2 = \overline{AB}$
$\overline{AB} = \overline{BC} = 2.81$

Relationships between circles

If two circles have radii which are in a ratio of $a:b$, then the following ratios are also true for the circles.

The diameters are also in the ratio of $a:b$.
The circumferences are also in the ratio $a:b$.
The areas are in the ratio $a^2:b^2$, or the ratio of the areas is the square of the ratios of the radii.

SKILL 3.14 Apply the theorems pertaining to the measures of inscribed angles and angles formed by chords, secants, and tangents.

<u>Angles with their vertices on the circle:</u>

An inscribed angle is an angle whose vertex is on the circle. Such an angle could be formed by two chords, two diameters, two secants, or a secant and a tangent. An inscribed angle has one arc of the circle in its interior. The measure of the inscribed angle is one-half the measure of this intercepted arc. If two inscribed angles intercept the same arc, the two angles are congruent (i.e. their measures are equal). If an inscribed angle intercepts an entire semicircle, the angle is a right angle.

<u>Angles with their vertices in a circle's interior:</u>

When two chords intersect inside a circle, two sets of vertical angles are formed. Each set of vertical angles intercepts two arcs which are across from each other. The measure of an angle formed by two chords in a circle is equal to one-half the sum of the angle intercepted by the angle and the arc intercepted by its vertical angle.

<u>Angles with their vertices in a circle's exterior:</u>

If an angle has its vertex outside of the circle and each side of the circle intersects the circle, then the angle contains two different arcs. The measure of the angle is equal to one-half the difference of the two arcs.

Examples:
1.

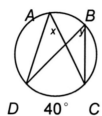

Find x and y.
arc DC = 40°

$m\angle DAC = \frac{1}{2}(40) = 20°$ $\angle DAC$ and $\angle DBC$ are both

$m\angle DBC = \frac{1}{2}(40) = 20°$ inscribed angles, so each one

$x = 20°$ and $y = 20°$ has a measure equal to one-half

the measure of arc DC.

SKILL 3.15 Identify basic geometric constructions (e.g., bisecting angles or line segments, constructing parallels or perpendiculars).

A geometric construction is a drawing made using only a compass and straightedge. A construction consists of only segments, arcs, and points.

Construct a line segment congruent to another line segment.

The easiest construction to make is to duplicate a given line segment. Given segment AB, construct a segment equal in length to segment AB by following these steps.

1. Place a point anywhere in the plane to anchor the duplicate segment. Call this point S.

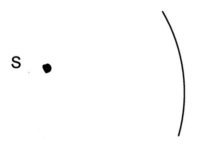

2. Open the compass to match the length of segment *AB*. Keeping the compass rigid, swing an arc from S.

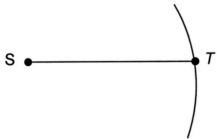

3. Draw a segment from S to any point on the arc. This segment will be the same length as *AB*.

Samples:

Construct segments congruent to the given segments.

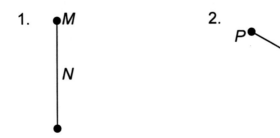

Construct an angle congruent to a given angle.

To construct an angle congruent to a given angle such as angle TAP follow these steps.

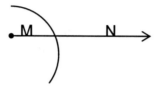

1. Draw ray *MN* using a straightedge. This ray will be one side of the duplicate angle.

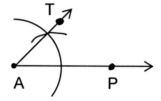

2. Using the compass, draw an arc of any radius with its central at the
   A. vertex Draw an arc of the same radius with center M.

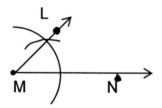

3. Use the point where the arc intercepts ray *AP* to draw another arc that intercepts the intersection of the arc and ray *AT*. Swing an arc of the same radius from the intersection point on ray *MN*.

4. Connect M and the point of intersection of the two arcs to form angle *LMN* which will be congruent to angle *TAP*.

Bisect an angle.

To bisect a given angle such as angle *FUZ*, follow these steps.

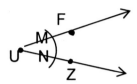

1. Swing an arc of any length with its center at point U. This arc will intersect rays *UF* and *UZ* at M and N.

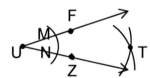

2. Open the compass to any length and swing one arc from point M and another arc of the same radius from point N. These arcs will intersect in the interior of angle *FUZ* at point T.

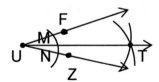

3. Connect U and T for the ray which bisects angle *FUZ*. Ray *UT* is the angle bisector of angle *FUZ*.

Given a point on a line, construct a perpendicular to the line through the point.

Given a line such as line $\overline{AB}$ and a point K on the line, follow these steps to construct a perpendicular line to line l through K.

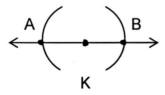

1. Swing an arc of any radius from point K so that it intersects line $\overline{AB}$ in two points, A and B.

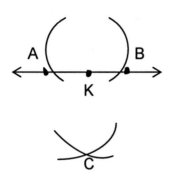

2. Open the compass to any length and swing one arc from B and another from A so that the two arcs intersect at point C.

3. Connect K and C to form line KC which is perpendicular to line $\overline{AB}$.

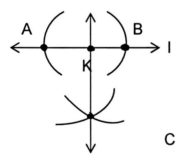

Construct the perpendicular to a given line through a point from a given point not on the line.

Given a line such as line l and a point P not on l, follow these steps to construct a perpendicular line to l that passes through P.

1. Swing an arc of any radius from P so that the arc intersects line l in two points A and B.

2. Open the compass to any length and swing two arcs of the same radius, one from A and the other from B. These two arcs will intersect at a new point K.

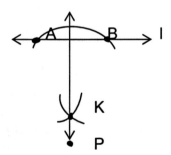

3. Connect K and P to form a line perpendicular to line l which passes through P.

Construct the perpendicular bisector of a line segment of a given line segment.

Given a line segment with two endpoints such as A and B, follow these steps to construct the line which both bisects and is perpendicular to the line given segment.

1. Swing an arc of any radius from point A. Swing another arc of the same radius from B. The arcs will intersect at two points. Label these points C and D.

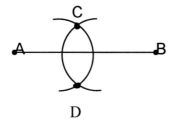

2. Connect C and D to form the perpendicular bisector of segment AB

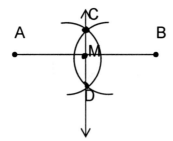

3. The point M where line $\overline{CD}$ and segment $\overline{AB}$ intersect is the midpoint of segment $\overline{AB}$.

Construct a parallel to a given line through a given point not on the line.

Given a point such as P and a line such as line m, follow these steps to construct the single line which passes through P and is also parallel to line m.

1. Place a point A anywhere on line m.

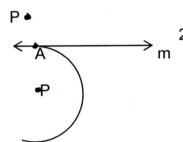
2. Open the compass to the distance Between A and P. Using P as the center, swing a long arc that passes through A.

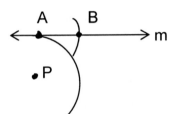
3. With the compass still open to the same length, swing an arc from A that intersects line m at a point labeled B.

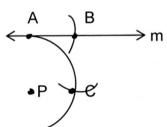
4. Using point B as the center, swing a new arc with the same radius as the other two arcs so that is intersects the arc from P at a new point C.

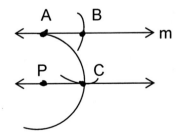
5. Connect P and C to obtain a line parallel to obtain a line parallel to line m.

Construct the tangent to a circle at a given point.

Given a circle with center O and a point on the circle such as P, construct the line tangent to the circle at P by constructing the line perpendicular to the radius drawn to P. If a line is perpendicular to a radius, the line will be tangent to the circle. For constructing a tangent to a circle, follow these steps.

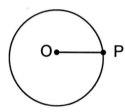

1. Draw the radius from O to P.

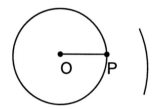

2. Open the compass from point P to point O. Use this radius to swing an arc from P to the exterior of the circle.

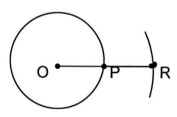

3. Put the straightedge on the radius and extend the segment to the arc forming segment OR. Note that P is the midpoint of OR.

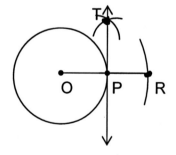

4. Open the compass to any length longer than the radius of this circle. Swing an arc of this radius from each endpoint of segment OR. Label the point where these two arcs intersect T.

5. Connect the point P and T to form the tangent line to circle O at point P.

SKILL 3.16 Identify the converse, inverse, and contrapositive of a conditional statement.

Conditional: If p, then q
  p is the hypothesis. q is the conclusion.

Inverse: If ~p, then ~q. Negate both the hypothesis (If not p, then not q) and the conclusion from the original conditional.

Converse : If q, then p. Reverse the 2 clauses. The original hypothesis becomes the conclusion. The original conclusion then becomes the new hypothesis.

Contrapositive: If ~q, then ~p. Reverse the 2 clauses. The If not q, then not p original hypothesis becomes the conclusion. The original conclusion then becomes the new hypothesis. THEN negate both the new hypothesis and the new conclusion.

Example: Given the **conditional**:

If an angle has 60°, then it is an acute angle.

Its **inverse**, in the form "If ~p, then ~q", would be:

If an angle doesn't have 60°, then it is not an acute angle.

NOTICE that the inverse is not true, even though the conditional statement was true.

Its **converse**, in the form "If q, then p", would be:

If an angle is an acute angle, then it has 60°.

NOTICE that the converse is not true, even though the conditional statement was true.

Its **contrapositive**, in the form "If q, then p", would be:

If an angle isn't an acute angle, then it doesn't have 60°.

NOTICE that the contrapositive is true, assuming the original conditional statement was true.

TIP: If you are asked to pick a statement that is logically equivalent to a given conditional, look for the contra-positive. The inverse and converse are not always logically equivalent to every conditional. The contra-positive is ALWAYS logically equivalent.

Find the inverse, converse and contrapositive of the following conditional statement. Also determine if each of the 4 statements is true or false.

Conditional: If $x = 5$, then $x^2 - 25 = 0$.   TRUE
Inverse: If $x \neq 5$, then $x^2 - 25 \neq 0$.   FALSE, x could be $^-5$
Converse: If $x^2 - 25 = 0$, then $x = 5$.   FALSE, x could be $^-5$
Contrapositive: If $x^2 - 25 \neq 0$, then $x \neq 5$.   TRUE

Conditional: If $x = 5$, then $6x = 30$.   TRUE
Inverse: If $x \neq 5$, then $6x \neq 30$.   TRUE
Converse: If $6x = 30$, then $x = 5$.   TRUE
Contrapositive: If $6x \neq 30$, then $x \neq 5$.   TRUE

Sometimes, as in this example, all 4 statements can be logically equivalent; however, the only statement that will always be logically equivalent to the original conditional is the contrapositive.

SKILL 3.17 Identify valid conclusions from given statements.

Conditional statements can be diagrammed using a **Venn diagram**. A diagram can be drawn with one figure inside another figure. The inner figure represents the hypothesis. The outer figure represents the conclusion. If the hypothesis is taken to be true, then you are located inside the inner figure. If you are located in the inner figure then you are also inside the outer figure, so that proves the conclusion is true. Sometimes that conclusion can then be used as the hypothesis for another conditional, which can result in a second conclusion.

Suppose that these statements were given to you, and you are asked to try to reach a conclusion. The statements are:

All swimmers are athletes.
All athletes are scholars.

In "if-then" form, these would be:
If you are a swimmer, then you are an athlete.
If you are an athlete, then you are a scholar.

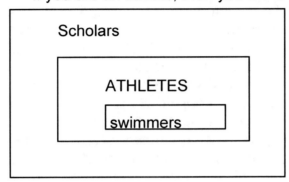

Clearly, if you are a swimmer, then you are also an athlete. This includes you in the group of scholars.

Suppose that these statements were given to you, and you are asked to try to reach a conclusion. The statements are:

All swimmers are athletes.
All wrestlers are athletes.

In "if-then" form, these would be:
If you are a swimmer, then you are an athlete.
If you are a wrestler, then you are an athlete.

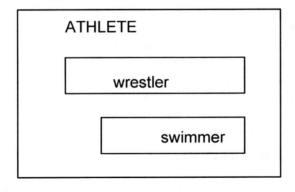

Clearly, if you are a swimmer or a wrestler, then you are also an athlete. This does NOT allow you to come to any other conclusions.

A swimmer may or may NOT also be a wrestler. Therefore, NO CONCLUSION IS POSSIBLE.

Suppose that these statements were given to you, and you are asked to try to reach a conclusion. The statements are:

All rectangles are parallelograms.
Quadrilateral ABCD is not a parallelogram.

In "if-then" form, the first statement would be:
If a figure is a rectangle, then it is also a parallelogram.

Note that the second statement is the negation of the conclusion of statement one. Remember also that the contrapositive is logically equivalent to a given conditional. That is, **"If ~ q, then ~ p"**. Since" ABCD is NOT a parallelogram " is like saying **"If ~ q,"** then you can come to the conclusion **"then ~ p"**. Therefore, the conclusion is ABCD is not a rectangle. Looking at the Venn diagram below, if all rectangles are parallelograms, then rectangles are included as part of the parallelograms. Since quadrilateral ABCD is not a parallelogram, that it is excluded from anywhere inside the parallelogram box. This allows you to conclude that ABCD can not be a rectangle either.

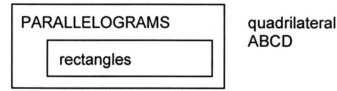

Try These:

What conclusion, if any, can be reached? Assume each statement is true, regardless of any personal beliefs.

1. If the Red Sox win the World Series, I will die.
   I died.

2. If an angle's measure is between 0° and 90°, then the angle is acute. Angle B is not acute.

3. Students who do well in geometry will succeed in college.
   Annie is doing extremely well in geometry.

4. Left-handed people are witty and charming.
   You are left-handed.

SKILL 3.18 Classify examples of reasoning processes as inductive or deductive.

Inductive thinking is the process of finding a pattern from a group of examples. That pattern is the conclusion that this set of examples seemed to indicate. It may be a correct conclusion or it may be an incorrect conclusion because other examples may not follow the predicted pattern.

Deductive thinking is the process of arriving at a conclusion based on other statements that are all known to be true, such as theorems, axioms postulates, or postulates. Conclusions found by deductive thinking based on true statements will **always** be true.

Examples :

Suppose:

> On Monday Mr.Peterson eats breakfast at McDonalds.
> On Tuesday Mr.Peterson eats breakfast at McDonalds.
> On Wednesday Mr.Peterson eats breakfast at McDonalds.
> On Thursday Mr.Peterson eats breakfast at McDonalds again.

Conclusion: On Friday Mr. Peterson will eat breakfast at McDonalds again.

This is a conclusion based on inductive reasoning. Based on several days observations, you conclude that Mr. Peterson will eat at McDonalds. This may or may not be true, but it is a conclusion arrived at by inductive thinking.

SKILL 3.19 Determine the surface area and volume of prisms, pyramids, cylinders, cones, and spheres.

Use the formulas to find the volume and surface area.

FIGURE	VOLUME	TOTAL SURFACE AREA
Right Cylinder	$\pi r^2 h$	$2\pi rh + 2\pi r^2$
Right Cone	$\dfrac{\pi r^2 h}{3}$	$\pi r \sqrt{r^2 + h^2} + \pi r^2$
Sphere	$\dfrac{4}{3}\pi r^3$	$4\pi r^2$
Rectangular Solid	LWH	$2LW + 2WH + 2LH$

Note: $\sqrt{r^2 + h^2}$ is equal to the slant height of the cone.

Sample problem:

1. Given the figure below, find the volume and surface area.

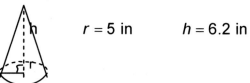

$r = 5$ in $h = 6.2$ in

Volume $= \dfrac{\pi r^2 h}{3}$ First write the formula.

$\dfrac{1}{3}\pi(5^2)(6.2)$ Then substitute.

162.31562 cubic inches Finally solve the problem.

Surface area $= \pi r \sqrt{r^2 + h^2} + \pi r^2$ First write the formula.

$\pi 5\sqrt{5^2 + 6.2^2} + \pi 5^2$ Then substitute.
203.652 square inches Compute.

Note: volume is always given in cubic units and area is always given in square units.

FIGURE	LATERAL AREA	TOTAL AREA	VOLUME
Right prism	sum of area of lateral faces (rectangles)	lateral area plus 2 times the area of base	area of base times Height
Regular pyramid	sum of area of lateral faces (triangles)	lateral area plus area of base	1/3 times the area of the base times the height

Find the total area of the given figure:

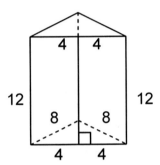

1. Since this is a triangular prism, first find the area of the bases.

2. Find the area of each rectangular lateral face.

3. Add the areas together.

$A = \dfrac{1}{2}bh$ $A = LW$ 1. write formula

$8^2 = 4^2 + h^2$
$h = 6.928$

2. find the height of the base triangle

$A = \dfrac{1}{2}(8)(6.928)$ $A = (8)(12)$ 3. substitute known values

$A = 27.713$ sq. units $A = 96$ sq. units   4. compute

Total Area $= 2(27.713) + 3(96)$
$= 343.426$ sq. units

Right circular cylinders and cones.

FIGURE	VOLUME	TOTAL SURFACE AREA	LATERAL AREA
Right Cylinder	$\pi r^2 h$	$2\pi rh + 2\pi r^2$	$2\pi rh$
Right Cone	$\dfrac{\pi r^2 h}{3}$	$\pi r \sqrt{r^2 + h^2} + \pi r^2$	$\pi r \sqrt{r^2 + h^2}$

Note: $\sqrt{r^2 + h^2}$ is equal to the slant height of the cone.

Sample problem:

1. A water company is trying to decide whether to use traditional cylindrical paper cups or to offer conical paper cups since both cost the same. The traditional cups are 8 cm wide and 14 cm high. The conical cups are 12 cm wide and 19 cm high. The company will use the cup that holds the most water.

1. Draw and label a sketch of each.

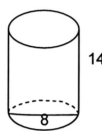
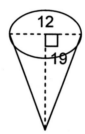

$V = \pi r^2 h$ $\quad\quad V = \dfrac{\pi r^2 h}{3}$ $\quad\quad$ 1. write formula

$V = \pi(4)^2(14)$ $\quad\quad V = \dfrac{1}{3}\pi(6)^2(19)$ $\quad\quad$ 2. substitute

$V = 703.717 \text{ cm}^3$ $\quad\quad V = 716.283 \text{ cm}^3$ $\quad\quad$ 3. solve

The choice should be the conical cup since its volume is more.

Spheres

FIGURE	VOLUME	TOTAL SURFACE AREA
Sphere	$\dfrac{4}{3}\pi r^3$	$4\pi r^2$

Sample problem:

1. How much material is needed to make a basketball that has a diameter of 15 inches? How much air is needed to fill the basketball?

Draw and label a sketch:

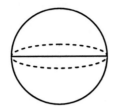   D=15 inches

Total surface area

$TSA = 4\pi r^2$

$= 4\pi(7.5)^2$

$= 706.858 \text{ in}^2$

Volume

$V = \dfrac{4}{3}\pi r^3$

$= \dfrac{4}{3}\pi(7.5)^3$

$= 1767.1459 \text{ in}^3$

1. write formula
2. substitute
3. solve

SKILL 3.20 Identify solids and their related nets.

The union of all points on a simple closed surface and all points in its interior form a space figure called a **solid**. The five regular solids, or **polyhedra**, are the cube, tetrahedron, octahedron, icosahedron, and dodecahedron. A **net** is a two-dimensional figure that can be cut out and folded up to make a three-dimensional solid. Below are models of the five regular solids with their corresponding face polygons and nets.

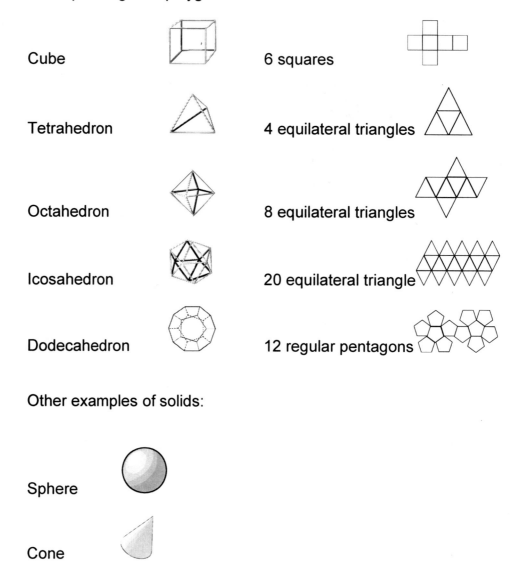

Cube — 6 squares

Tetrahedron — 4 equilateral triangles

Octahedron — 8 equilateral triangles

Icosahedron — 20 equilateral triangle

Dodecahedron — 12 regular pentagons

Other examples of solids:

Sphere

Cone

COMPETENCY 4.0 KNOWLEDGE OF GEOMETRY FROM AN ALGEBRAIC PERSPECTIVE

SKILL 4.1 Solve distance and midpoint problems involving two points, a point and a line, two lines, and two parallel lines.

Apply the distance formula (two points).

The key to applying the distance formula is to understand the problem before beginning.

$$D = \sqrt{(x_2 - x_1)^2 + (y_2 - y_1)^2}$$

Sample Problem:

1. Find the perimeter of a figure with vertices at $(4,5)$, $(^-4,6)$ and $(^-5,^-8)$.

The figure being described is a triangle. Therefore, the distance for all three sides must be found. Carefully, identify all three sides before beginning.

Side 1 = $(4,5)$ to $(^-4,6)$
Side 2 = $(^-4,6)$ to $(^-5,^-8)$
Side 3 = $(^-5,^-8)$ to $(4,5)$

$$D_1 = \sqrt{(^-4-4)^2 + (6-5)^2} = \sqrt{65}$$

$$D_2 = \sqrt{((^-5-(^-4))^2 + (^-8-6)^2} = \sqrt{197}$$

$$D_3 = \sqrt{((4-(^-5))^2 + (5-(^-8))^2} = \sqrt{250} \text{ or } 5\sqrt{10}$$

$$\text{Perimeter} = \sqrt{65} + \sqrt{197} + 5\sqrt{10}$$

Apply the formula for midpoint.

Midpoint Definition:

If a line segment has endpoints of (x_1, y_1) and (x_2, y_2), then the midpoint can be found using:

$$\left(\frac{x_1 + x_2}{2}, \frac{y_1 + y_2}{2} \right)$$

Sample problems:

1. Find the center of a circle with a diameter whose endpoints are $(3,7)$ and $(^-4, ^-5)$.

$$\text{Midpoint} = \left(\frac{3 + (^-4)}{2}, \frac{7 + (^-5)}{2} \right)$$

$$\text{Midpoint} = \left(\frac{^-1}{2}, 1 \right)$$

2. Find the midpoint given the two points $\left(5, 8\sqrt{6}\right)$ and $\left(9, ^-4\sqrt{6}\right)$.

$$\text{Midpoint} = \left(\frac{5+9}{2}, \frac{8\sqrt{6} + (^-4\sqrt{6})}{2} \right)$$

$$\text{Midpoint} = \left(7, 2\sqrt{6}\right)$$

Determine the distance between a point and a line.

In order to accomplish the task of finding the distance from a given point to another given line the perpendicular line that intersects the point and line must be drawn and the equation of the other line written. From this information the point of intersection can be found. This point and the original point are used in the distance formula given below:

$$D = \sqrt{(x_2 - x_1)^2 + (y_2 - y_1)^2}$$

Sample Problem:

1. Given the point $(^-4, 3)$ and the line $y = 4x + 2$, find the distance from the point to the line.

$y = 4x + 2$	1. Find the slope of the given line by solving for y.
$y = 4x + 2$	2. The slope is 4/1, the perpendicular line will have a slope of $^-1/4$.
$y = \left(^-1/4\right)x + b$	3. Use the new slope and the given point to find the equation of the perpendicular line.
$3 = \left(^-1/4\right)\left(^-4\right) + b$	4. Substitute $(^-4, 3)$ into the equation.
$3 = 1 + b$	5. Solve.
$2 = b$	6. Given the value for b, write the equation of the perpendicular line.
$y = \left(^-1/4\right)x + 2$	7. Write in standard form.
$x + 4y = 8$	8. Use both equations to solve the point of intersection.
$^-4x + y = 2$	
$x + 4y = 8$	9. Multiply the bottom row by 4.
$^-4x + y = 2$	
$4x + 16y = 32$	
$17y = 34$	10. Solve.
$y = 2$	
$y = 4x + 2$	11. Substitute to find the x value.
$2 = 4x + 2$	12. Solve.
$x = 0$	

(0,2) is the point of intersection. Use this point on the original line and the original point to calculate the distance between them.

$D = \sqrt{(x_2 - x_1)^2 + (y_2 - y_1)^2}$	where points are (0,2) and (-4,3).
$D = \sqrt{(^-4 - 0)^2 + (3 - 2)^2}$	1. Substitute.
$D = \sqrt{(16) + (1)}$	2. Simplify.
$D = \sqrt{17}$	

MATHEMATICS 6-12

Determine the distance between two parallel lines.

The distance between two parallel lines, such as line AB and line CD as shown below is the line segment RS, the perpendicular between the two parallels.

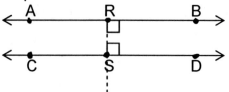

Sample Problem:

Given the geometric figure below, find the distance between the two parallel sides AB and CD.

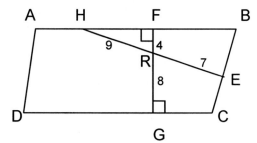

The distance FG is 12 units.

SKILL 4.2 Identify the directrix, foci, vertices, axes, and asymptotes of a conic section where appropriate.

PARABOLAS-A parabola is a set of all points in a plane that are equidistant from a fixed point (focus) and a line (directrix).

FORM OF EQUATION $y = a(x-h)^2 + k$ $x = a(y-k)^2 + h$

IDENTIFICATION x^2 term, y not squared y^2 term, x not squared

SKETCH OF GRAPH

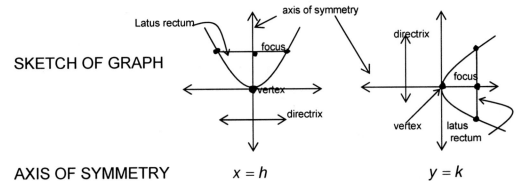

AXIS OF SYMMETRY $x = h$ $y = k$
  -A line through the vertex and focus upon which the parabola is symmetric.

VERTEX (h,k) (h,k)

  -The point where the parabola intersects the axis of symmetry.

FOCUS $(h, k + 1/4a)$ $(h + 1/4a, k)$

DIRECTRIX $y = k - 1/4a$ $x = h - 1/4a$

DIRECTION OF OPENING   up if $a > 0$, down if $a < 0$ right if $a > 0$, left if $a < 0$

LENGTH OF LATUS RECTUM $|1/a|$ $|1/a|$

  -A chord through the focus, perpendicular to the axis of symmetry, with endpoints on the parabola.

Sample Problem:

1. Find all identifying features of $y = {}^-3x^2 + 6x - 1$.

First, the equation must be put into the general form $y = a(x-h)^2 + k$.

$y = {}^-3x^2 + 6x - 1$ 1. Begin by completing the square.
$= {}^-3(x^2 - 2x + 1) - 1 + 3$
$= {}^-3(x-1)^2 + 2$ 2. Using the general form of the equation to identify known variables.

$a = {}^-3 \quad h = 1 \quad k = 2$
axis of symmetry: $x = 1$
vertex: $(1, 2)$
focus: $(1, 1\frac{1}{4})$
directrix: $y = 2\frac{3}{4}$
direction of opening: down since $a < 0$
length of latus rectum: $1/3$

ELLIPSE

FORM OF EQUATION $\dfrac{(x-h)^2}{a^2}+\dfrac{(y-k)^2}{b^2}=1$ $\dfrac{(x-h)^2}{b^2}+\dfrac{(y-k)^2}{a^2}=1$

(for ellipses where $a^2 > b^2$).

where $b^2 = a^2 - c^2$ where $b^2 = a^2 - c^2$

IDENTIFICATION   horizontal major axis   vertical major axis

SKETCH

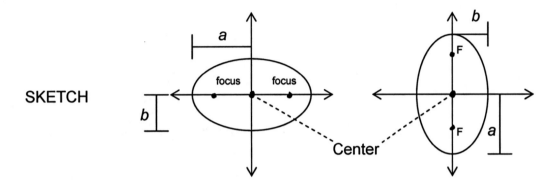

CENTER	(h,k)	(h,k)
FOCI	$(h \pm c, k)$	$(h, k \pm c)$
MAJOR AXIS LENGTH	$2a$	$2a$
MINOR AXIS LENGTH	$2b$	$2b$

Sample Problem:

Find all identifying features of the ellipse $2x^2 + y^2 - 4x + 8y - 6 = 0$.

First, begin by writing the equation in standard form for an ellipse.

$2x^2 + y^2 - 4x + 8y - 6 = 0$ 1. Complete the square for each variable.

$2(x^2 - 2x + 1) + (y^2 + 8y + 16) = 6 + 2(1) + 16$

$2(x-1)^2 + (y+4)^2 = 24$ 2. Divide both sides by 24.

$\dfrac{(x-1)^2}{12} + \dfrac{(y+4)^2}{24} = 1$

    3. Now the equation is in standard form.

Identify known variables: $h = 1$ $k = {}^-4$ $a = \sqrt{24}$ or $2\sqrt{6}$
$b = \sqrt{12}$ or $2\sqrt{3}$ $c = 2\sqrt{3}$

Identification: vertical major axis
Center: $(1, {}^-4)$
Foci: $(1, {}^-4 \pm 2\sqrt{3})$
Major axis: $4\sqrt{6}$
Minor axis: $4\sqrt{3}$

HYPERBOLA

FORM OF EQUATION

$$\frac{(x-h)^2}{a^2} - \frac{(y-k)^2}{b^2} = 1 \qquad \frac{(y-k)^2}{a^2} - \frac{(x-h)^2}{b^2} = 1$$

where $c^2 = a^2 + b^2$ — where $c^2 = a^2 + b^2$

IDENTIFICATION — horizontal transverse axis (y^2 is negative) — vertical transverse axis (x^2 is negative)

SKETCH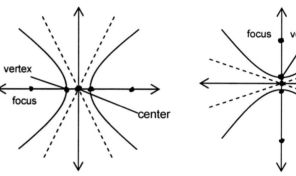

SLOPE OF ASYMPTOTES	$\pm(b/a)$	$\pm(a/b)$
TRANSVERSE AXIS (endpoints are vertices of the hyperbola and go through the center)	$2a$ -on y axis	$2a$ -on x axis
CONJUGATE AXIS (perpendicular to transverse axis at center)	$2b$, -on y axis	$2b$, -on x axis
CENTER	(h,k)	(h,k)
FOCI	$(h \pm c, k)$	$(h, k \pm c)$
VERTICES	$(h \pm a, k)$	$(h, k \pm a)$

Sample Problem:

Find all the identifying features of a hyperbola given its equation.

$$\frac{(x+3)^2}{4} - \frac{(y-4)^2}{16} = 1$$

Identify all known variables: $h = {}^-3 \quad k = 4 \quad a = 2 \quad b = 4 \quad c = 2\sqrt{5}$

Slope of asymptotes: $\pm 4/2$ or ± 2
Transverse axis: 4 units long
Conjugate axis: 8 units long
Center: $({}^-3, 4)$
Foci: $({}^-3 \pm 2\sqrt{5}, 4)$
Vertices: $({}^-1, 4)$ and $({}^-5, 4)$

SKILL 4.3 Determine the center and the radius of a circle given its equation, and identify the graph.

The equation of a circle with its center at (h,k) and a radius r units is:

$$(x-h)^2 + (y-k)^2 = r^2$$

Sample Problem:

1. Given the equation $x^2 + y^2 = 9$, find the center and the radius of the circle. Then graph the equation.

First, writing the equation in standard circle form gives:

$$(x-0)^2 + (y-0)^2 = 3^2$$

therefore, the center is (0,0) and the radius is 3 units.

Sketch the circle:

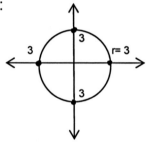

2. Given the equation $x^2 + y^2 - 3x + 8y - 20 = 0$, find the center and the radius. Then graph the circle.

First, write the equation in standard circle form by completing the square for both variables.

$x^2 + y^2 - 3x + 8y - 20 = 0$ 1. Complete the squares.
$(x^2 - 3x + 9/4) + (y^2 + 8y + 16) = 20 + 9/4 + 16$
$(x - 3/2)^2 + (y + 4)^2 = 153/4$

The center is $(3/2, {}^-4)$ and the radius is $\dfrac{\sqrt{153}}{2}$ or $\dfrac{3\sqrt{17}}{2}$.

Graph the circle.

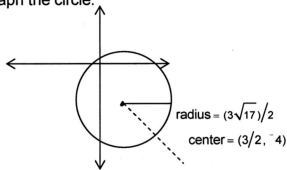

radius = $(3\sqrt{17})/2$
center = $(3/2, {}^-4)$

To write the equation given the center and the radius use the standard form of the equation of a circle:

$$(x - h)^2 + (y - k)^2 = r^2$$

Sample problems:
Given the center and radius, write the equation of the circle.
1. Center $({}^-1, 4)$; radius 11

    $(x - h)^2 + (y - k)^2 = r^2$ 1. Write standard equation.
    $(x - ({}^-1))^2 + (y - (4))^2 = 11^2$ 2. Substitute.
    $(x + 1)^2 + (y - 4)^2 = 121$ 3. Simplify.

2. Center $(\sqrt{3}, {}^-1/2)$; radius = $5\sqrt{2}$

    $(x - h)^2 + (y - k)^2 = r^2$ 1. Write standard equation.

    $(x - \sqrt{3})^2 + (y - ({}^-1/2))^2 = (5\sqrt{2})^2$ 2. Substitute.
    $(x - \sqrt{3})^2 + (y + 1/2)^2 = 50$ 3. Simplify.

SKILL 4.4 Identify the equation of a conic section, given the appropriate information.

Conic sections result from the intersection of a cone and a plane. The three main types of conics are parabolas, ellipses, and hyperbolas.

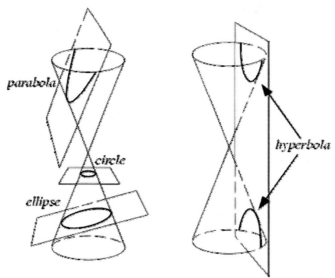

The general equation for a conic section is:

$$Ax^2 + Bxy + Cy^2 + Dx + Ey + F = 0$$

The value of $B^2 - 4AC$ determines the type of conic. If $B^2 - 4AC$ is less than zero the curve is an ellipse or a circle. If equal to zero, the curve is a parabola. If greater than zero, the curve is a hyperbola.

A **parabola** is the set of all points in a plane equidistant from a fixed point called the focus (F) and a fixed line called the directrix. The vertex is the point halfway between the focus and the directrix that lies on the parabola.

The equation of a parabola with focus $(h, k + p)$, directrix $y = k - p$, and vertex (h, k) is

$$(x - h)^2 = 4p(y - k)$$

If $p > 0$ the parabola opens up. If $p < 0$ the parabola opens down.

Example:

Find the equation of a parabola with vertex (1,2) and focus (1,3).

k + p = 2	Definition of focus.
1+ p = 2	Substitute for k (vertex value).
p = 1	Solve for p.
$(x - 1)^2 = 4(p)(y - 2)$	Equation of parabola
$(x - 1)^2 = 4(y - 2)$	Substitute for p.

An **ellipse** is the set of all points in a plane the sum of whose distances from two fixed points called foci is equal. The vertices of an ellipse are the two points farthest from the center. The major axis joins the vertices. The general equation of an ellipse with center located at (h, k), vertices (h±a, k), foci (h±c, k)

$$\frac{(x-h)^2}{a^2}+\frac{(y-k)^2}{b^2}=1$$ where a ≥ b > 0 and $a^2 - b^2 = c^2$

The distance between the foci is 2c. The length of the major axis is 2a.

Example:

Find the equation of an ellipse with foci (4, -4), (6, -4) and vertices (3, -2), (7, -2).

The length of the major axis joining the vertices (3, -2), (7, -2) is 4, so a = 2. The distance between the foci (4, -4), (6, -4) is 2, so c = 1. Therefore, $b^2 = a^2 - c^2 = 3$. The center of the ellipse is (5, -2).

Substituting the resolved values into the equation of an ellipse yields...

$$\frac{(x-5)^2}{4}+\frac{(y+2)^2}{3}=1$$

A **hyperbola** is the set of all points in a plane the difference of whose distances from two fixed points called foci is equal. The vertices of a hyperbola are the two points where the curve makes its sharpest turns located on the major axis (the line through the foci). The general equation of a hyperbola centered at (h, k), foci (h±c, k), vertices (h±a, k) and asymptotes $y - k = \pm\frac{b}{a}(x-h)$

$$\frac{(x-h)^2}{a^2} - \frac{(y-k)^2}{b^2} = 1 \quad \text{where } a^2 + b^2 = c^2$$

The distance between the foci is 2c. The distance between the vertices is 2a.

Example:

Find the equation of a hyperbola with foci (1, 3) and (7, 3) and vertices (2, 3) and (6, 3).

The distance between the foci is 6, thus c = 3. The distance between the vertices is 4, thus a = 2. Therefore, $b^2 = c^2 - a^2 = 3^2 - 2^2 = 5$. The hyperbola is centered at (4, 3).

Substituting the resolved values into the equation of a hyperbola yields...

$$\frac{(x-4)^2}{4} - \frac{(y-3)^2}{5} = 1$$

SKILL 4.5 Use translations, rotations, dilations, or reflections on a coordinate plane to identify the images of geometric objects under such transformations.

Example:

Plot the given ordered pairs on a coordinate plane and join them in the given order, then join the first and last points.

(-3, -2), (3, -2), (5, -4), (5, -6), (2, -4), (-2, -4), (-5, -6), (-5, -4)

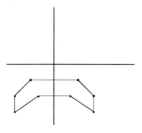

Increase all y-coordinates by 6.
(-3, 4), (3, 4), (5, 2), (5, 0), (2, 2), (-2, 2), (-5, 0), (-5, 2)

Plot the points and join them to form a second figure.

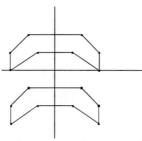

A figure on a coordinate plane can be translated by changing the ordered pairs.

A **transformation matrix** defines how to map points from one coordinate space into another coordinate space. The matrix used to accomplish two-dimensional transformations is described mathematically by a 3-by-3 matrix.

Example:
A point transformed by a 3-by-3 matrix

$$[x \; y \; 1] \begin{pmatrix} a & b & u \\ c & d & v \\ t_x & t_y & w \end{pmatrix} x = [x' \; y' \; 1]$$

A 3-by-3 matrix transforms a point (x, y) into a point (x', y') by means of the following equations:

$$x' = ax + cy + t_x$$

$$y' = bx + dy + t_y$$

Another type of transformation is **dilation**. Dilation is a transformation that "shrinks" or "makes it bigger."

Example:

Using dilation to transform a diagram.

Starting with a triangle whose center of dilation is point P,

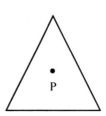

we dilate the lengths of the sides by the same factor to create a new triangle.

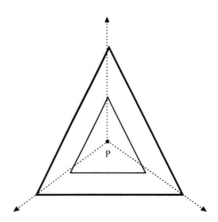

COMPETENCY 5.0 KNOWLEDGE OF TRIGONOMETRY

SKILL 5.1 Identify equations of graphs of circular/trigonometric functions and their inverses.

Unlike trigonometric identities that are true for all values of the defined variable, trigonometric equations are true for some, but not all, of the values of the variable. Most often trigonometric equations are solved for values between 0 and 360 degrees or 0 and 2π radians.

Some algebraic operation, such as squaring both sides of an equation, will give you extraneous answers. You must remember to check all solutions to be sure that they work.

Sample problems:

1. Solve: $\cos x = 1 - \sin x$ if $0 \leq x < 360$ degrees.

$\cos^2 x = (1 - \sin x)^2$	1. square both sides
$1 - \sin^2 x = 1 - 2\sin x + \sin^2 x$	2. substitute
$0 = {}^-2\sin x + 2\sin^2 x$	3. set = to 0
$0 = 2\sin x({}^-1 + \sin x)$	4. factor
$2\sin x = 0 \quad {}^-1 + \sin x = 0$	5. set each factor = 0
$\sin x = 0 \qquad \sin x = 1$	6. solve for $\sin x$
$x = 0$ or $180 \quad x = 90$	7. find value of sin at x

   The solutions appear to be 0, 90 and 180. Remember to check each solution and you will find that 180 does not give you a true equation. Therefore, the only solutions are 0 and 90 degrees.

2. Solve: $\cos^2 x = \sin^2 x$ if $0 \leq x < 2\pi$

$\cos^2 x = 1 - \cos^2 x$	1. substitute
$2\cos^2 x = 1$	2. simplify
$\cos^2 x = \dfrac{1}{2}$	3. divide by 2
$\sqrt{\cos^2 x} = \pm\sqrt{\dfrac{1}{2}}$	4. take square root
$\cos x = \dfrac{\pm\sqrt{2}}{2}$	5. rationalize denominator
$x = \dfrac{\pi}{4}, \dfrac{3\pi}{4}, \dfrac{5\pi}{4}, \dfrac{7\pi}{4}$	

SKILL 5.2 Solve problems involving circular/trigonometric function identities.

Prove circular/trigonometric function identities.

Given 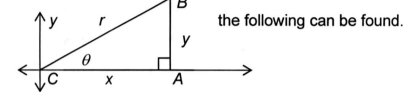 the following can be found.

Trigonometric Functions:

$$\sin\theta = \frac{y}{r} \qquad \csc\theta = \frac{r}{y}$$

$$\cos\theta = \frac{x}{r} \qquad \sec\theta = \frac{r}{x}$$

$$\tan\theta = \frac{y}{x} \qquad \cot\theta = \frac{x}{y}$$

Sample problem:

1. Prove that $\sec\theta = \dfrac{1}{\cos\theta}$.

$\sec\theta = \dfrac{1}{\frac{x}{r}}$ Substitution definition of cosine.

$\sec\theta = \dfrac{1 \times r}{\frac{x}{r} \times r}$ Multiply by $\dfrac{r}{r}$.

$\sec\theta = \dfrac{r}{x}$ Substitution.

$\sec\theta = \sec\theta$ Substitute definition of $\dfrac{r}{x}$.

$\sec\theta = \dfrac{1}{\cos\theta}$ Substitute.

2. Prove that $\sin^2 + \cos^2 = 1$.

$\left(\dfrac{y}{r}\right)^2 + \left(\dfrac{x}{r}\right)^2 = 1$ Substitute definitions of sin and cos.

$\dfrac{y^2 + x^2}{r^2} = 1$ $x^2 + y^2 = r^2$ Pythagorean formula.

$\dfrac{r^2}{r^2} = 1$ Simplify.

$1 = 1$ Substitute.

$\sin^2\theta + \cos^2\theta = 1$

Practice problems: Prove each identity.

1. $\cot\theta = \dfrac{\cos\theta}{\sin\theta}$ 2. $1 + \cot^2\theta = \csc^2\theta$

Apply basic circular/trigonometric function identities.

There are two methods that may be used to prove trigonometric identities. One method is to choose one side of the equation and manipulate it until it equals the other side. The other method is to replace expressions on both sides of the equation with equivalent expressions until both sides are equal.

The Reciprocal Identities

$\sin x = \dfrac{1}{\csc x}$ $\sin x \csc x = 1$ $\csc x = \dfrac{1}{\sin x}$

$\cos x = \dfrac{1}{\sec x}$ $\cos x \sec x = 1$ $\sec x = \dfrac{1}{\cos x}$

$\tan x = \dfrac{1}{\cot x}$ $\tan x \cot x = 1$ $\cot x = \dfrac{1}{\tan x}$

$\tan x = \dfrac{\sin x}{\cos x}$ $\cot x = \dfrac{\cos x}{\sin x}$

The Pythagorean Identities

$\sin^2 x + \cos^2 x = 1$ $1 + \tan^2 x = \sec^2 x$ $1 + \cot^2 x = \csc^2 x$

Sample problems:

1. Prove that $\cot x + \tan x = (\csc x)(\sec x)$.

$\dfrac{\cos x}{\sin x} + \dfrac{\sin x}{\cos x}$ Reciprocal identities.

$\dfrac{\cos^2 x + \sin^2 x}{\sin x \cos x}$ Common denominator.

$\dfrac{1}{\sin x \cos x}$ Pythagorean identity.

$\dfrac{1}{\sin x} \times \dfrac{1}{\cos x}$

$\csc x(\sec x) = \csc x(\sec x)$ Reciprocal identity, therefore,
$\cot x + \tan x = \csc x(\sec x)$

2. Prove that $\dfrac{\cos^2 \theta}{1 + 2\sin\theta + \sin^2 \theta} = \dfrac{\sec\theta - \tan\theta}{\sec\theta + \tan\theta}$.

$\dfrac{1 - \sin^2 \theta}{(1+\sin\theta)(1+\sin\theta)} = \dfrac{\sec\theta - \tan\theta}{\sec\theta + \tan\theta}$ Pythagorean identity

factor denominator.

$\dfrac{1 - \sin^2 \theta}{(1+\sin\theta)(1+\sin\theta)} = \dfrac{\dfrac{1}{\cos\theta} - \dfrac{\sin\theta}{\cos\theta}}{\dfrac{1}{\cos\theta} + \dfrac{\sin\theta}{\cos\theta}}$ Reciprocal identities.

$\dfrac{(1-\sin\theta)(1+\sin\theta)}{(1+\sin\theta)(1+\sin\theta)} = \dfrac{\dfrac{1-\sin\theta}{\cos\theta}(\cos\theta)}{\dfrac{1+\sin\theta}{\cos\theta}(\cos\theta)}$ Factor $1-\sin^2\theta$.

Multiply by $\dfrac{\cos\theta}{\cos\theta}$.

$\dfrac{1-\sin\theta}{1+\sin\theta} = \dfrac{1-\sin\theta}{1+\sin\theta}$ Simplify.

$\dfrac{\cos^2\theta}{1+2\sin\theta+\sin^2\theta} = \dfrac{\sec\theta-\tan\theta}{\sec\theta+\tan\theta}$

SKILL 5.3   Interpret the graphs of trigonometric functions (e.g., amplitude, period, phase shift).

The trigonometric functions sine, cosine, and tangent are periodic functions. The values of periodic functions repeat on regular intervals. Period, amplitude, and phase shift are key properties of periodic functions that can be determined by observation of the graph.

The **period** of a function is the smallest domain containing the complete cycle of the function. For example, the period of a sine or cosine function is the distance between the peaks of the graph.

The **amplitude** of a function is half the distance between the maximum and minimum values of the function.

Phase shift is the amount of horizontal displacement of a function from its original position.

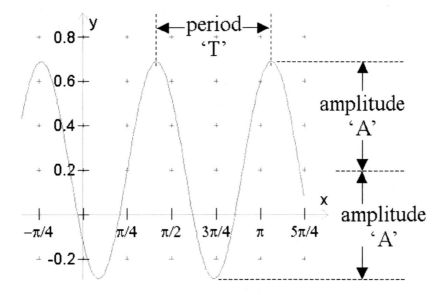

Properties of the graphs of basic trigonometric functions.

Function	Period	Amplitude
y = sin x	2π radians	1
y = cos x	2π radians	1
y = tan x	π radians	undefined

Below are the graphs of the basic trigonometric functions, (a) y = sin x; (b) y = cos x; and (c) y = tan x.

A)   B)   C)

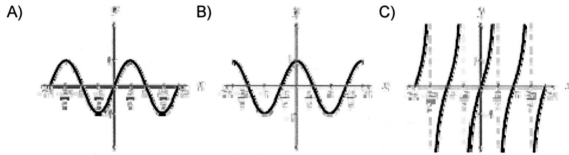

Note that the phase shift of trigonometric graphs is the horizontal distance displacement of the curve from these basic functions.

SKILL 5.4 Solve real-world problems involving triangles using the law of sines or the law of cosines.

Apply the law of sines.

Definition: For any triangle ABC, where a, b, and c are the lengths of the sides opposite angles A, B, and C respectively.

$$\frac{\sin A}{a} = \frac{\sin B}{b} = \frac{\sin C}{c}$$

Sample problem:

1. An inlet is 140 feet wide. The lines of sight from each bank to an approaching ship are 79 degrees and 58 degrees. What are the distances from each bank to the ship?

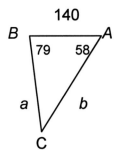

1. Draw and label a sketch.

2. The missing angle is
   $180 - (79 + 58) = 43$
   degrees.

$$\frac{\sin A}{a} = \frac{\sin B}{b} = \frac{\sin C}{c}$$

3. Write formula.

Side opposite 79 degree angle:

$$\frac{\sin 79}{b} = \frac{\sin 43}{140}$$

4. Substitute.

$$b = \frac{140(.9816)}{.6820}$$

5. Solve.

$b \approx 201.501$ feet

Side opposite 58 degree angle:

$$\frac{\sin 58}{a} = \frac{\sin 43}{140}$$

6. Substitute.

$$a = \frac{140(.848)}{.6820}$$

7. Solve.

$a \approx 174.076$ feet

Apply the law of cosines.

Definition: For any triangle ABC, when given two sides and the included angle, the other side can be found using one of the formulas below:

$$a^2 = b^2 + c^2 - (2bc)\cos A$$
$$b^2 = a^2 + c^2 - (2ac)\cos B$$
$$c^2 = a^2 + b^2 - (2ab)\cos C$$

Similarly, when given three sides of a triangle, the included angles can be found using the derivation:

$$\cos A = \frac{b^2 + c^2 - a^2}{2bc}$$
$$\cos B = \frac{a^2 + c^2 - b^2}{2ac}$$
$$\cos C = \frac{a^2 + b^2 - c^2}{2ab}$$

Sample problem:

1. Solve triangle ABC, if angle $B = 87.5°$, $a = 12.3$, and $c = 23.2$. (Compute to the nearest tenth).

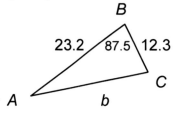

1. Draw and label a sketch.

Find side b.

$b^2 = a^2 + c^2 - (2ac)\cos B$ 2. Write the formula.

$b^2 = (12.3)^2 + (23.2)^2 - 2(12.3)(23.2)(\cos 87.5)$ 3. Substitute.

$b^2 = 664.636$

$b = 25.8$ (rounded)   4. Solve.

Use the law of sines to find angle A.

$\dfrac{\sin A}{a} = \dfrac{\sin B}{b}$ 1. Write formula.

$\dfrac{\sin A}{12.3} = \dfrac{\sin 87.5}{25.8} = \dfrac{12.29}{25.8}$ 2. Substitute.

$\sin A = 0.47629$ 3. Solve.

Angle $A = 28.4$

Therefore, angle $C = 180 - (87.5 + 28.4)$
$= 64.1$

2. Solve triangle ABC if $a = 15$, $b = 21$, and $c = 18$. (Round to the nearest tenth).

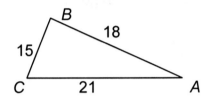

1. Draw and label a sketch.

Find angle A.

$\cos A = \dfrac{b^2 + c^2 - a^2}{2bc}$ 2. Write formula.

$\cos A = \dfrac{21^2 + 18^2 - 15^2}{2(21)(18)}$ 3. Substitute.

$\cos A = 0.714$ 4. Solve.

Angle $A = 44.4$

Find angle B.

$\cos B = \dfrac{a^2 + c^2 - b^2}{2ac}$ 5. Write formula.

$\cos B = \dfrac{15^2 + 18^2 - 21^2}{2(15)(18)}$ 6. Substitute.

$\cos B = 0.2$ 7. Solve.

Angle $B = 78.5$

Therefore, angle $C = 180 - (44.4 + 78.5)$
$= 57.1$

SKILL 5.5 **Use tangent, sine, and cosine ratios to solve right triangle problems.**

Use the basic trigonometric ratios of sine, cosine and tangent to solve for the missing sides of right triangles when given at least one of the acute angles.

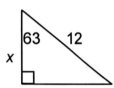

In the triangle ABC, an acute angle of 63 degrees and the length of the hypotenuse (12). The missing side is the one adjacent to the given angle.

The appropriate trigonometric ratio to use would be cosine since we are looking for the adjacent side and we have the length of the hypotenuse.

$$\text{Cos} x = \frac{\text{adjacent}}{\text{hypotenuse}}$$ 1. Write formula.

$$\text{Cos} 63 = \frac{x}{12}$$ 2. Substitute known values.

$$0.454 = \frac{x}{12}$$ 3. Solve.

$$x = 5.448$$

Sample problem:

1. Find the missing side.

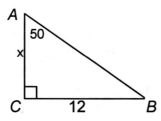

1. Identify the known values. Angle $A = 50$ degrees and the side opposite the given angle is 12. The missing side is the adjacent leg.

2. The information suggests the use of the tangent function

$\tan A = \dfrac{\text{opposite}}{\text{adjacent}}$ 3. Write the function.

$\tan 50 = \dfrac{12}{x}$ 4. Substitute.

$1.192 = \dfrac{12}{x}$ 5. Solve.

$x(1.192) = 12$

$x = 10.069$

Remember that since angle A and angle B are complimentary, then angle $B = 90 - 50$ or 40 degrees.

Using this information we could have solved for the same side only this time it is the leg opposite from angle B.

$\tan B = \dfrac{\text{opposite}}{\text{adjacent}}$ 1. Write the formula.

$\tan 40 = \dfrac{x}{12}$ 2. Substitute.

$12(.839) = x$ 3. Solve.

$10.069 \approx x$

Now that the two sides of the triangle are known, the third side can be found using the Pythagorean Theorem.

COMPETENCY 6.0 KNOWLEDGE OF STATISTICS

SKILL 6.1 Interpret graphical data involving measures of location (i.e., percentiles, stanines, quartiles).

Percentiles divide data into 100 equal parts. A person whose score falls in the 65th percentile has outperformed 65 percent of all those who took the test. This does not mean that the score was 65 percent out of 100 nor does it mean that 65 percent of the questions answered were correct. It means that the grade was higher than 65 percent of all those who took the test.

Stanine "standard nine" scores combine the understandability of percentages with the properties of the normal curve of probability. Stanines divide the bell curve into nine sections, the largest of which stretches from the 40th to the 60th percentile and is the "Fifth Stanine" (the average of taking into account error possibilities).

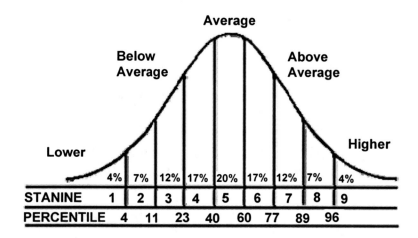

Quartiles divide the data into 4 parts. First find the median of the data set (Q2), then find the median of the upper (Q3) and lower (Q1) halves of the data set. If there are an odd number of values in the data set, include the median value in both halves when finding quartile values. For example, given the data set: {1, 4, 9, 16, 25, 36, 49, 64, 81} first find the median value, which is 25 this is the second quartile. Since there are an odd number of values in the data set (9), we include the median in both halves. To find the quartile values, we much find the medians of: {1, 4, 9, 16, 25} and {25, 36, 49, 64, 81}. Since each of these subsets had an odd number of elements (5), we use the middle value. Thus the first quartile value is 9 and the third quartile value is 49. If the data set had an even number of elements, average the middle two values. The quartile values are always either one of the data points, or exactly half way between two data points.

Sample problem:

1. Given the following set of data, find the percentile of the score 104.

   70, 72, 82, 83, 84, 87, 100, 104, 108, 109, 110, 115

Solution: Find the percentage of scores below 104.

7/12 of the scores are less than 104. This is 58.333%; therefore, the score of 104 is in the 58th percentile.

2. Find the first, second and third quartile for the data listed.

   6, 7, 8, 9, 10, 12, 13, 14, 15, 16, 18, 23, 24, 25, 27, 29, 30, 33, 34, 37

Quartile 1: The 1st Quartile is the median of the lower half of the data set, which is 11.

Quartile 2: The median of the data set is the 2nd Quartile, which is 17.

Quartile 3: The 3rd Quartile is the median of the upper half of the data set, which is 28.

SKILL 6.2   Compute the mean, median, and mode of a set of data.

Mean, median and mode are three measures of central tendency. The **mean** is the average of the data items. The **median** is found by putting the data items in order from smallest to largest and selecting the item in the middle (or the average of the two items in the middle). The **mode** is the most frequently occurring item.

Example:

Find the mean, median, and mode of the test score listed below:

85	77	65
92	90	54
88	85	70
75	80	69
85	88	60
72	74	95

Mean (X) = sum of all scores ÷ number of scores
    = 78

Median = put numbers in order from smallest to largest. Pick middle number.

54, 60, 65, 69, 70, 72, 74, 75, 77, 80, 85, 85, 85, 88, 88, 90, 92, 95

both in middle

Therefore, median is average of two numbers in the middle or 78.5

Mode = most frequent number
    = 85

SKILL 6.3 Determine whether the mean, the median, or the mode is the most appropriate measure of central tendency in a given situation.

Different situations require different information. If we examine the circumstances under which an ice cream store owner may use statistics collected in the store, we find different uses for different information.

Over a 7-day period, the store owner collected data on the ice cream flavors sold. He found the mean number of scoops sold was 174 per day. The most frequently sold flavor was vanilla. This information was useful in determining how much ice cream to order in all and in what amounts for each flavor.

In the case of the ice cream store, the median and range had little business value for the owner. Consider the set of test scores from a math class: 0, 16, 19, 65, 65, 65, 68, 69, 70, 72, 73, 73, 75, 78, 80, 85, 88, and 92. The mean is 64.06 and the median is 71. Since there are only three scores less than the mean out of the eighteen scores, the median (71) would be a more descriptive score.

Retail store owners may be most concerned with the most common dress size so they may order more of that size than any other.

SKILL 6.4 Interpret the ranges, variances, and standard deviations for ungrouped data.

An understanding of the definitions is important in determining the validity and uses of statistical data. All definitions and applications in this section apply to ungrouped data.

Data item: each piece of data is represented by the letter X.

Mean: the average of all data represented by the symbol $\overline{X}$.

Range: difference between the highest and lowest value of data items.

Sum of the Squares: sum of the squares of the differences between each item and the mean. $Sx^2 = (X - \overline{X})^2$

Variance: the sum of the squares quantity divided by the number of items.
(the lower case Greek letter sigma squared (σ^2) represents variance).

$$\frac{Sx^2}{N} = \sigma^2$$

The larger the value of the variance the larger the spread

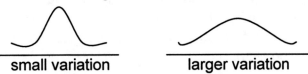

small variation   larger variation

Standard Deviation: the square root of the variance. The lower case Greek letter sigma (σ) is used to represent standard deviation. $\sigma = \sqrt{\sigma^2}$

Most statistical calculators have standard deviation keys on them and should be used when asked to calculate statistical functions. It is important to become familiar with the calculator and the location of the keys needed.

Sample Problem:

Given the ungrouped data below, calculate the mean, range, standard deviation and the variance.

15 22 28 25 34 38
18 25 30 33 19 23

Mean ($\overline{X}$) = 25.8333333
Range: 38 − 15 = 23
standard deviation (σ) = 6.99137
Variance (σ^2) = 48.87879

SKILL 6.5 **Interpret information from bar, line, picto-, and circle graphs; stem-and-leaf and scatter plots; and box-and-whisker graphs.**

Basic statistical concepts can be applied without computations. For example, inferences can be drawn from a graph or statistical data. A bar graph could display which grade level collected the most money. Student test scores would enable the teacher to determine which units need to be remediated.

To make a **bar graph** or a **pictograph**, determine the scale to be used for the graph. Then determine the length of each bar on the graph or determine the number of pictures needed to represent each item of information. Be sure to include an explanation of the scale in the legend.

Example: A class had the following grades:
4 A's, 9 B's, 8 C's, 1 D, 3 F's.

Graph these on a bar graph and a pictograph.

Pictograph

Grade	Number of Students
A	☺☺☺☺
B	☺☺☺☺☺☺☺☺☺
C	☺☺☺☺☺☺☺☺
D	☺
F	☺☺☺

Bar graph

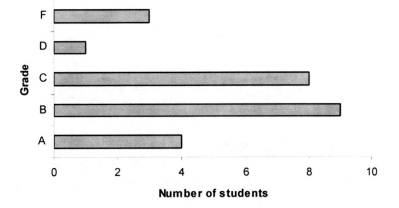

To read a bar graph or a pictograph, read the explanation of the scale that was used in the legend. Compare the length of each bar with the dimensions on the axes and calculate the value each bar represents. On a pictograph count the number of pictures used in the chart and calculate the value of all the pictures.

To make a **line graph**, determine appropriate scales for both the vertical and horizontal axes (based on the information to be graphed). Describe what each axis represents and mark the scale periodically on each axis. Graph the individual points of the graph and connect the points on the graph from left to right.

Example: Graph the following information using a line graph.

The number of National Merit finalists/school year

	90-'91	91-'92	92-'93	93-'94	94-'95	95-'96
Central	3	5	1	4	6	8
Wilson	4	2	3	2	3	2

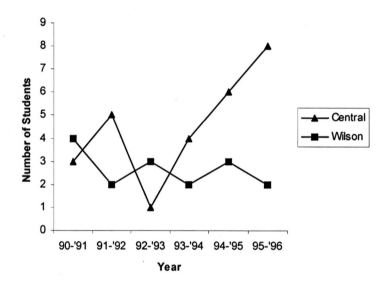

To make a **circle graph**, total all the information that is to be included on the graph. Determine the central angle to be used for each sector of the graph using the following formula:

$$\frac{\text{information}}{\text{total information}} \times 360° = \text{degrees in central} \sphericalangle$$

Lay out the central angles to these sizes, label each section and include its percent.

Example: Graph this information on a circle graph:

Monthly expenses:

Rent, $400
Food, $150
Utilities, $75
Clothes, $75
Church, $100
Misc., $200

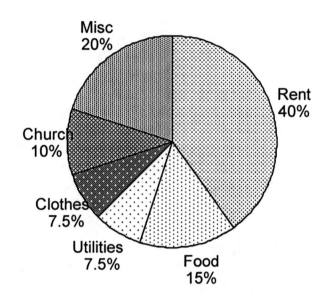

To read a circle graph, find the total of the amounts represented on the entire circle graph. To determine the actual amount that each sector of the graph represents, multiply the percent in a sector times the total amount number.

Scatter plots compare two characteristics of the same group of things or people and usually consist of a large body of data. They show how much one variable is affected by another. The relationship between the two variables is their **correlation**. The closer the data points come to making a straight line when plotted, the closer the correlation.

Stem and leaf plots are visually similar to line plots. The **stems** are the digits in the greatest place value of the data values, and the **leaves** are the digits in the next greatest place values. Stem and leaf plots are best suited for small sets of data and are especially useful for comparing two sets of data. The following is an example using test scores:

4	9
5	4 9
6	1 2 3 4 6 7 8 8
7	0 3 4 6 6 6 7 7 7 8 8 8 8
8	3 5 5 7 8
9	0 0 3 4 5
10	0 0

Histograms are used to summarize information from large sets of data that can be naturally grouped into intervals. The vertical axis indicates **frequency** (the number of times any particular data value occurs), and the horizontal axis indicates data values or ranges of data values. The number of data values in any interval is the **frequency of the interval**.

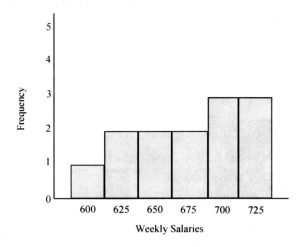

SKILL 6.6 **Interpret problems involving basic statistical concepts such as sampling, experimental design, correlation, and linear regression.**

Random sampling supplies every combination of items from the frame, or stratum, as a known probability of occurring. A large body of statistical theory quantifies the risk and thus enables an appropriate sample size to be chosen.

Systematic sampling selects items in the frame according to the kth sample. The first item is chosen to be the rth, where r is a random integer in the range $1,...,k-1$.

There are three stages to Cluster or Area sampling: the target population is divided into many regional clusters (groups); a few clusters are randomly selected for study; a few subjects are randomly chosen from within a cluster.

Convenience sampling is the method of choosing items arbitrarily and in an unstructured manner from the frame.

* * *

Correlation is a measure of association between two variables. It varies from -1 to 1, with 0 being a random relationship, 1 being a perfect positive linear relationship, and -1 being a perfect negative linear relationship.

The **correlation coefficient** (r) is used to describe the strength of the association between the variables and the direction of the association.

Example:

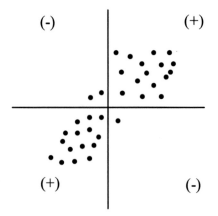

Horizontal and vertical lines are drawn through the point of averages which is the point on the averages of the x and y values. This divides the scatter plot into four quadrants. If a point is in the lower left quadrant, the product of two negatives is positive; in the upper right, the product of two positives is positive. The positive quadrants are depicted with the positive sign (+). In the two remaining quadrants (upper left and lower right), the product of a negative and a positive is negative. The negative quadrants are depicted with the negative sign (-). If r is positive, then there are more points in the positive quadrants and if r is negative, then there are more points in the two negative quadrants.

Regression is a form of statistical analysis used to predict a dependent variable (y) from values of an independent variable (x). A regression equation is derived from a known set of data.

The simplest regression analysis models the relationship between two variables using the following equation: $y = a + bx$, where y is the dependent variable and x is the independent variable. This simple equation denotes a linear relationship between x and y. This form would be appropriate if, when you plotted a graph of x and y, you tended to see the points roughly form along a straight line.

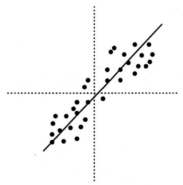

The line can then be used to make predictions.

If all of the data points fell on the line, there would be a perfect correlation ($r = 1.0$) between the x and y data points. These cases represent the best scenarios for prediction. A positive or negative r value represents how y varies with x. When r is positive, y increases as x increases. When r is negative y decreases as x increases.

A **linear regression** equation is of the form: $Y = a + bX$.

Example:

A teacher wanted to determine how a practice test influenced a student's performance on the actual test. The practice test grade and the subsequent actual test grade for each student are given in the table below:

Practice Test (x)	Actual Test (y)
94	98
95	94
92	95
87	89
82	85
80	78
75	73
65	67
50	45
20	40

We determine the equation for the linear regression line to be $y = 14.650 + 0.834x$.

A new student comes into the class and scores 78 on the practice test. Based on the equation obtained above, what would the teacher predict this student would get on the actual test?

$$y = 14.650 + 0.834(78)$$
$$y = 14.650 + 65.052$$
$$y = 80$$

COMPETENCY 7.0 KNOWLEDGE OF PROBABILITY

SKILL 7.1 Determine probabilities of dependent or independent events.

Dependent events occur when the probability of the second event depends on the outcome of the first event. For example, consider the two events (A) it is sunny on Saturday and (B) you go to the beach. If you intend to go to the beach on Saturday, rain or shine, then A and B may be independent. If however, you plan to go to the beach only if it is sunny, then A and B may be dependent. In this situation, the probability of event B will change depending on the outcome of event A.

Suppose you have a pair of dice, one red and one green. If you roll a three on the red die and then roll a four on the green die, we can see that these events do not depend on the other. The total probability of the two independent events can be found by multiplying the separate probabilities.

$$P(A \text{ and } B) = P(A) \times P(B)$$
$$= 1/6 \times 1/6$$
$$= 1/36$$

Many times, however, events are not independent. Suppose a jar contains 12 red marbles and 8 blue marbles. If you randomly pick a red marble, replace it and then randomly pick again, the probability of picking a red marble the second time remains the same. However, if you pick a red marble, and then pick again without replacing the first red marble, the second pick becomes dependent upon the first pick.

$$P(\text{Red and Red}) \text{ with replacement} = P(\text{Red}) \times P(\text{Red})$$
$$= 12/20 \times 12/20$$
$$= 9/25$$
$$P(\text{Red and Red}) \text{ without replacement} = P(\text{Red}) \times P(\text{Red})$$
$$= 12/20 \times 11/19$$
$$= 33/95$$

SKILL 7.2 Predict odds of a given outcome.

Odds are defined as the ratio of the number of favorable outcomes to the number of unfavorable outcomes. The sum of the favorable outcomes and the unfavorable outcomes should always equal the total possible outcomes.

For example, given a bag of 12 red and 7 green marbles compute the odds of randomly selecting a red marble.

$$\text{Odds of red} = \frac{12}{19}$$

$$\text{Odds of not getting red} = \frac{7}{19}$$

In the case of flipping a coin, it is equally likely that a head or a tail will be tossed. The odds of tossing a head are 1:1. This is called even odds.

SKILL 7.3 Identify an appropriate sample space for an experiment.

In probability, the **sample space** is a list of all possible outcomes of an experiment. For example, the sample space of tossing two coins is the set {HH, HT, TT, TH}, the sample space of rolling a six-sided die is the set {1, 2, 3, 4, 5, 6}, and the sample space of measuring the height of students in a class is the set of all real numbers {R}.

When conducting experiments with a large number of possible outcomes it is important to determine the size of the sample space. The size of the sample space can be determined by using the fundamental counting principle and the rules of combinations and permutations.

The **fundamental counting principle** states that if there are m possible outcomes for one task and n possible outcomes of another, there are
($m \times n$) possible outcomes of the two tasks together.

A **permutation** is the number of possible arrangements of items, without repetition, where order of selection is important.

A **combination** is the number of possible arrangements, without repetition, where order of selection is not important.

Permutations and combinations are covered in detail in Skill 9.4.

Examples:

1. Find the size of the sample space of rolling two six-sided die and flipping two coins.

    Solution:

    List the possible outcomes of each event:

    each dice: {1, 2, 3, 4, 5, 6}
    each coin: {Heads, Tails}

    Apply the fundamental counting principle:
    size of sample space = 6 x 6 x 2 x 2 = 144

2. Find the size of the sample space of selecting three playing cards at random from a standard fifty-two card deck.

    Solution:

    Use the rule of combination –
    $$_{52}C_3 = \frac{52!}{(52-3)!3!} = 22100$$

SKILL 7.4   Make predictions that are based on relative frequency of an event.

The absolute probability of some events cannot be determined. For instance, one cannot assume the probability of winning a tennis match is ½ because, in general, winning and losing are not equally likely. In such cases, past results of similar events can be used to help predict future outcomes. The **relative frequency** of an event is the number of times an event has occurred divided by the number of attempts.

Relative frequency = $\frac{\text{number of successful trials}}{\text{total number of trials}}$

For example, if a weighted coin flipped 50 times lands on heads 40 times and tails 10 times, the relative frequency of heads is 40/50 = 4/5. Thus, one can predict that if the coin is flipped 100 times, it will land on heads 80 times.

Example:

Two tennis players, John and David, have played each other 20 times.

John has won 15 of the previous matches and David has won 5.
(a) Estimate the probability that David will win the next match.
(b) Estimate the probability that John will win the next 3 matches.

Solution:

(a) David has won 5 out of 20 matches. Thus, the relative frequency of David winning is 5/20 or ¼. We can estimate that the probability of David winning the next match is ¼.

(b) John has won 15 out of 20 matches. The relative frequency of John winning is 15/20 or ¾. We can estimate that the probability of John winning a future match is ¾. Thus, the probability that John will win the next three matches is ¾ x ¾ x ¾ = 27/64.

SKILL 7.5 **Determine probabilities using counting procedures, tables, tree diagrams, and formulas for permutations and combinations.**

The Addition Principle of Counting states:

If A and B are events, then $n(A \text{ or } B) = n(A) + n(B) - n(A \cap B)$.

Example:

In how many ways can you select a black card or a Jack from an ordinary deck of playing cards?

Let B denote the set of black cards and let J denote the set of Jacks. Then, $n(B) = 26, n(J) = 4, n(B \cap J) = 2$ and

$$n(B \text{ or } J) = n(B) + n(J) - n(B \cap A)$$
$$= 26 + 4 - 2$$
$$= 28.$$

The Addition Principle of Counting for Mutually Exclusive Events states:

If A and B are mutually exclusive events, then
$n(A \text{ or } B) = n(A) + n(B)$.

Example:

A travel agency offers 40 possible trips: 14 to Asia, 16 to Europe and 10 to South America. In how many ways can you select a trip to Asia or Europe through this agency?

Let A denote trips to Asia and let E denote trips to Europe. Then, $A \cap E = \varnothing$ and
$$n(A \text{ or } E) = 14 + 16 = 30.$$

Therefore, the number of ways you can select a trip to Asia or Europe is 30.

The Multiplication Principle of Counting for Dependent Events states:

Let A be a set of outcomes of Stage 1 and B a set of outcomes of Stage 2. Then the number of ways $n(A \text{ and } B)$, that A and B can occur in a two-stage experiment is given by:
$$n(A \text{ and } B) = n(A)n(B|A),$$

where $n(B|A)$ denotes the number of ways B can occur given that A has already occurred.

Example:

How many ways from an ordinary deck of 52 cards can two Jacks be drawn in succession if the first card is drawn but not replaced in the deck and then the second card is drawn?

This is a two-stage experiment for which we wish to compute $n(A \text{ and } B)$, where A is the set of outcomes for which a Jack is obtained on the first draw and B is the set of outcomes for which a Jack is obtained on the second draw.

If the first card drawn is a Jack, then there are only three remaining Jacks left to choose from on the second draw. Thus, drawing two cards without replacement means the events A and B are dependent.

$$n(A \text{ and } B) = n(A)n(B|A) = 4 \cdot 3 = 12$$

The Multiplication Principle of Counting for Independent Events states:

Let A be a set of outcomes of Stage 1 and B a set of outcomes of Stage 2. If A and B are independent events then the number of ways $n(A and B)$, that A and B can occur in a two-stage experiment is given by:

$$n(A and B) = n(A)n(B).$$

Example:

How many six-letter code "words" can be formed if repetition of letters is not allowed?

Since these are code words, a word does not have to look like a word; for example, abcdef could be a code word. Since we must choose a first letter *and* a second letter *and* a third letter *and* a fourth letter *and* a fifth letter *and* a sixth letter, this experiment has six stages.

Since repetition is not allowed there are 26 choices for the first letter; 25 for the second; 24 for the third; 23 for the fourth; 22 for the fifth; and 21 for the sixth. Therefore, we have:

n(six-letter code words without repetition of letters)

$$= 26 \cdot 25 \cdot 24 \cdot 23 \cdot 22 \cdot 21$$

$$= 165,765,600$$

A **Bernoulli trial** is an experiment whose outcome is random and can be either of two possible outcomes, called "success" or "failure." Tossing a coin would be an example of a Bernoulli trial. We make the outcomes into a random variable by assigning the number 0 to one outcome and the number 1 to the other outcome. Traditionally, the "1" outcome is considered the "success" and the "0" outcome is considered the "failure." The probability of success is represented by p, with the probability of failure being $1-p$, or q.

Bernoulli trials can be applied to any real-life situation in which there are just two possible outcomes. For example, concerning the birth of a child, the only two possible outcomes for the sex of the child are male or female.

The **binomial distribution** is a sequence of probabilities with each probability corresponding to the likelihood of a particular event occurring. It is called a binomial distribution because each trial has precisely two possible outcomes. An **event** is defined as a sequence of Bernoulli trials that has within it a specific number of successes. The order of success is not important.

Note: There are two parameters to consider in a binomial distribution:

1. p = the probability of a success
2. n = the number of Bernoulli trials (i.e., the length of the sequence).

Example:

Toss a coin two times. Each toss is a Bernoulli trial as discussed above. Consider heads to be success. One event is one sequence of two coin tosses. Order does not matter.

There are two possibilities for each coin toss. Therefore, there are four (2·2) possible subevents: 00, 01, 10, 11 (where 0 = tail and 1 = head).

According to the multiplication rule, each subevent has a probability of $\frac{1}{4}\left(\frac{1}{2}\cdot\frac{1}{2}\right)$.

One subevent has zero heads, so the event of zero heads in two tosses is:
$$p(h=0)=\frac{1}{4}.$$

Two subevents have one head, so the event of one head in two tosses is:
$$p(h=1)=\frac{2}{4}.$$

One subevent has two heads, so the event of two heads in two tosses is:
$$p(h=2)=\frac{1}{4}.$$

So the binomial distribution for two tosses of a fair coin is:
$$p(h=0)=\frac{1}{4},\ p(h=1)=\frac{2}{4},\ p(h=2)=\frac{1}{4}.$$

A **normal distribution** is the distribution associated with most sets of real-world data. It is frequently called a **bell curve**. A normal distribution has a **random variable** X with mean μ and variance σ^2.

Example:

Albert's Bagel Shop's morning customer load follows a normal distribution, with **mean** (average) 50 and **standard deviation** 10. The standard deviation is the measure of the variation in the distribution. Determine the probability that the number of customers tomorrow will be less than 42.

First convert the raw score to a **z-score**. A z-score is a measure of the distance in standard deviations of a sample from the mean.

The z-score = $\dfrac{X_i - \bar{X}}{s} = \dfrac{42 - 50}{10} = \dfrac{-8}{10} = -.8$

Next, use a table to find the probability corresponding to the z-score. The table gives us .2881. Since our raw score is negative, we subtract the table value from .5.

$$.5 - .2881 = .2119$$

We can conclude that $P(x < 42) = .2119$. This means that there is about a 21% chance that there will be fewer than 42 customers tomorrow morning.

Example:

The scores on Mr. Rogers' statistics exam follow a normal distribution with mean 85 and standard deviation 5. A student is wondering what the probability is that she will score between a 90 and a 95 on her exam.

We wish to compute $P(90 < x < 95)$.

Compute the z-scores for each raw score.

$$\frac{90-85}{5} = \frac{5}{5} = 1 \text{ and } \frac{95-85}{5} = \frac{10}{5} = 2.$$

Now we want $P(1 < z < 2)$.

Since we are looking for an occurrence between two values, we subtract:

$$P(1 < z < 2) = P(z < 2) - P(z < 1).$$

We use a table to get

$P(1 < z < 2) = .9772 - .8413 = .1359$. (Remember that since the z-scores are positive, we add .5 to each probability.)

We can then conclude that there is a 13.6% chance that the student will score between a 90 and a 95 on her exam.

COMPETENCY 8.0 KNOWLEDGE OF DISCRETE MATHEMATICS

SKILL 8.1 Find a specified term in an arithmetic sequence.

When given a set of numbers where the common difference between the terms is constant, use the following formula:

$$a_n = a_1 + (n-1)d$$

where a_1 = the first term
n = the n th term (general term)
d = the common difference

Sample problem:

1. Find the 8th term of the arithmetic sequence 5, 8, 11, 14, ...

$a_n = a_1 + (n-1)d$
$a_1 = 5$ Identify 1st term.
$d = 3$ Find d.
$a_8 = 5 + (8-1)3$ Substitute.
$a_8 = 26$

2. Given two terms of an arithmetic sequence find a and d.

$a_4 = 21 \quad a_6 = 32$
$a_n = a_1 + (n-1)d$
$21 = a_1 + (4-1)d$
$32 = a_1 + (6-1)d$
$21 = a_1 + 3d$ Solve the system of equations.
$32 = a_1 + 5d$

$21 = a_1 + 3d$
$-32 = {}^-a_1 - 5d$ Multiply by $^-1$ and add the equations.
$\overline{{}^-11 = {}^-2d}$
$5.5 = d$
$21 = a_1 + 3(5.5)$ Substitute $d = 5.5$ into one of the equations.
$21 = a_1 + 16.5$
$a_1 = 4.5$

The sequence begins with 4.5 and has a common difference of 5.5 between numbers.

SKILL 8.2　Find a specified term in a geometric sequence.

When using geometric sequences consecutive numbers are compared to find the common ratio.

$$r = \frac{a_{n+1}}{a_n}$$

r = the common ratio
a_n = the nth term

The ratio is then used in the geometric sequence formula:

$$a_n = a_1 r^{n-1}$$

Sample problems:

1. Find the 8th term of the geometric sequence 2, 8, 32, 128 ...

$r = \dfrac{a_{n+1}}{a_n}$　　Use the common ratio formula to find r.

$r = \dfrac{8}{2} = 4$　　Substitute $a_n = 2$　　$a_{n+1} = 8$

$a_n = a_1 \times r^{n-1}$　　Use $r = 4$ to solve for the 8th term.
$a_8 = 2 \times 4^{8-1}$
$a_8 = 32768$

SKILL 8.3　Determine the sum of terms in an arithmetic or geometric progression.

The sums of terms in a progression is simply found by determining if it is an arithmetic or geometric sequence and then using the appropriate formula.

Sum of first n terms of an arithmetic sequence.

$$S_n = \frac{n}{2}(a_1 + a_n)$$

or

$$S_n = \frac{n}{2}\left[2a_1 + (n-1)d\right]$$

Sum of first n terms of a geometric sequence.

$$S_n = \frac{a_1(r^n - 1)}{r - 1}, r \neq 1$$

Sample Problems:

1. $\sum_{i=1}^{10}(2i + 2)$

   This means find the sum of the term beginning with the first term and ending with the 10th term of the sequence $a = 2i + 2$.

   $a_1 = 2(1) + 2 = 4$
   $a_{10} = 2(10) + 2 = 22$
   $S_n = \frac{n}{2}(a_1 + a_n)$
   $S_n = \frac{10}{2}(4 + 22)$
   $S_n = 130$

2. Find the sum of the first 6 terms in an arithmetic sequence if the first term is 2 and the common difference d, is -3.

   $n = 6 \qquad a_1 = 2 \qquad d = {}^-3$

   $S_n = \frac{n}{2}\left[2a_1 + (n-1)d\right]$

   $S_6 = \frac{6}{2}\left[2 \times 2 + (6-1){}^-3\right]$ Substitute known values.

   $S_6 = 3\left[4 + ({}^-15)\right]$ Solve.

   $S_6 = 3(-11) = -33$

3. Find $\sum_{i=1}^{5} 4 \times 2^i$

This means the sum of the first 5 terms where $a_i = a \times b^i$ and $r = b$.

$a_1 = 4 \times 2^1 = 8$ Identify a_1, r, n
$r = 2 \quad n = 5$

$S_n = \dfrac{a_1(r^n - 1)}{r - 1}$ Substitute a, r, n

$S_5 = \dfrac{8(2^5 - 1)}{2 - 1}$ Solve.

$S_5 = \dfrac{8(31)}{1} = 248$

Practice problems:

1. Find the sum of the first five terms of the sequence if $a = 7$ and $d = 4$.

2. $\sum_{i=1}^{7}(2i - 4)$

3. $\sum_{i=1}^{6} -3\left(\dfrac{2}{5}\right)^i$

SKILL 8.4   Solve problems involving permutations and combinations.

The difference between permutations and combinations is that in permutations all possible ways of writing an arrangement of objects are given while in a combination a given arrangement of objects is listed only once.

Given the set {1, 2, 3, 4}, list the arrangements of two numbers that can be written as a combination and as a permutation.

Combination	Permutation
12, 13, 14, 23, 24, 34	12, 21, 13, 31, 14, 41, 23, 32, 24, 42, 34, 43,
six ways	twelve ways

Using the formulas given below the same results can be found.

$$_nP_r = \frac{n!}{(n-r)!}$$

The notation $_nP_r$ is read "the number of permutations of n objects taken r at a time."

$$_4P_2 = \frac{4!}{(4-2)!}$$ Substitute known values.

$$_4P_2 = 12$$ Solve.

$$_nC_r = \frac{n!}{(n-r)!r!}$$ The number of combinations when r objects are selected from n objects.

$$_4C_2 = \frac{4!}{(4-2)!2!}$$ Substitute known values.

$$_4C_2 = 6$$ Solve.

SKILL 8.5 Evaluate matrix expressions involving sums, differences, and products.

A matrix is a square array of numbers called its entries or elements. The dimensions of a matrix are written as the number of rows (r) by the number of columns (r × c).

$$\begin{pmatrix} 1 & 2 & 3 \\ 4 & 5 & 6 \end{pmatrix}$$ is a 2 × 3 matrix (2 rows by 3 columns)

$$\begin{pmatrix} 1 & 2 \\ 3 & 4 \\ 5 & 6 \end{pmatrix}$$ is a 3 × 2 matrix (3 rows by 2 columns)

Associated with every square matrix is a number called the determinant.

Use these formulas to calculate determinants.

2×2 $\begin{pmatrix} a & b \\ c & d \end{pmatrix} = ad - bc$

3×3

$\begin{pmatrix} a_1 & b_1 & c_1 \\ a_2 & b_2 & c_2 \\ a_3 & b_3 & c_3 \end{pmatrix} = (a_1 b_2 c_3 + b_1 c_2 a_3 + c_1 a_2 b_3) - (a_3 b_2 c_1 + b_3 c_2 a_1 + c_3 a_2 b_1)$

This is found by repeating the first two columns and then using the diagonal lines to find the value of each expression as shown below:

$\begin{pmatrix} a_1^* & b_1^\circ & c_1^\bullet \\ a_2 & b_2^* & c_2^\circ \\ a_3 & b_3 & c_3^* \end{pmatrix} \begin{matrix} a_1 & b_1 \\ a_2^\bullet & b_2 \\ a_3^\circ & b_3^\bullet \end{matrix} = (a_1 b_2 c_3 + b_1 c_2 a_3 + c_1 a_2 b_3) - (a_3 b_2 c_1 + b_3 c_2 a_1 + c_3 a_2 b_1)$

Sample Problem:

1. Find the value of the determinant:

$\begin{pmatrix} 4 & ^-8 \\ 7 & 3 \end{pmatrix} = (4)(3) - (7)(^-8)$ Cross multiply and subtract.

$12 - (^-56) = 68$ Then simplify.

Sums and differences of matrices

Addition of matrices is accomplished by adding the corresponding elements of the two matrices. Subtraction is defined as the inverse of addition. In other words, change the sign on all the elements in the second matrix and add the two matrices.

Sample problems:
Find the sum or difference.

1. $\begin{pmatrix} 2 & 3 \\ ^-4 & 7 \\ 8 & ^-1 \end{pmatrix} + \begin{pmatrix} 8 & ^-1 \\ 2 & ^-1 \\ 3 & ^-2 \end{pmatrix} =$

$\begin{pmatrix} 2+8 & 3+(^-1) \\ ^-4+2 & 7+(^-1) \\ 8+3 & ^-1+(^-2) \end{pmatrix}$ Add corresponding elements.

$\begin{pmatrix} 10 & 2 \\ ^-2 & 6 \\ 11 & ^-3 \end{pmatrix}$ Simplify.

2. $\begin{pmatrix} 8 & ^-1 \\ 7 & 4 \end{pmatrix} - \begin{pmatrix} 3 & 6 \\ ^-5 & 1 \end{pmatrix} =$

$\begin{pmatrix} 8 & ^-1 \\ 7 & 4 \end{pmatrix} + \begin{pmatrix} ^-3 & ^-6 \\ 5 & ^-1 \end{pmatrix} =$

Change all of the signs in the second matrix and then add the two matrices.

$\begin{pmatrix} 8+(^-3) & ^-1+(^-6) \\ 7+5 & 4+(^-1) \end{pmatrix} =$ Simplify.

$\begin{pmatrix} 5 & ^-7 \\ 12 & 3 \end{pmatrix}$

Practice problems:

1. $\begin{pmatrix} 8 & ^-1 \\ 5 & 3 \end{pmatrix} + \begin{pmatrix} 3 & 8 \\ 6 & ^-2 \end{pmatrix} =$

2. $\begin{pmatrix} 3 & 7 \\ ^-4 & 12 \\ 0 & ^-5 \end{pmatrix} - \begin{pmatrix} 3 & 4 \\ 6 & ^-1 \\ ^-5 & ^-5 \end{pmatrix} =$

Products of a scalar and a matrix.

Scalar multiplication is the product of the scalar (the outside number) and each element inside the matrix.

Sample problem:

Given: $A = \begin{pmatrix} 4 & 0 \\ 3 & {}^-1 \end{pmatrix}$ Find 2A.

$$2A = 2\begin{pmatrix} 4 & 0 \\ 3 & {}^-1 \end{pmatrix}$$

$$\begin{pmatrix} 2\times 4 & 2\times 0 \\ 2\times 3 & 2\times {}^-1 \end{pmatrix}$$ Multiply each element in the matrix by the scalar.

$$\begin{pmatrix} 8 & 0 \\ 6 & {}^-2 \end{pmatrix}$$ Simplify.

Practice problems:

1. $\quad {}^-2\begin{pmatrix} 2 & 0 & 1 \\ {}^-1 & {}^-2 & 4 \end{pmatrix}$

2. $\quad 3\begin{pmatrix} 6 \\ 2 \\ 8 \end{pmatrix} + 4\begin{pmatrix} 0 \\ 7 \\ 2 \end{pmatrix}$

3. $\quad 2\begin{pmatrix} {}^-6 & 8 \\ {}^-2 & {}^-1 \\ 0 & 3 \end{pmatrix}$

Products of two matrices.

The product of two matrices can only be found if the number of columns in the first matrix is equal to the number of rows in the second matrix. Matrix multiplication is not necessarily commutative.

Sample problems:

1. Find the product AB if:

$$A = \begin{pmatrix} 2 & 3 & 0 \\ 1 & {}^-4 & {}^-2 \\ 0 & 1 & 1 \end{pmatrix} \qquad B = \begin{pmatrix} {}^-2 & 3 \\ 6 & {}^-1 \\ 0 & 2 \end{pmatrix}$$

$$3 \times 3 \qquad\qquad\qquad 3 \times 2$$

Note: Since the number of columns in the first matrix (3 × <u>3</u>) matches the number of rows (<u>3</u> ×2) this product is defined and can be found. The dimensions of the product will be equal to the number of rows in the first matrix (<u>3</u> ×3) by the number of columns in the second matrix (3 × <u>2</u>). The answer will be a 3 × 2 matrix.

$$AB = \begin{pmatrix} 2 & 3 & 0 \\ 1 & {}^-4 & {}^-2 \\ 0 & 1 & 1 \end{pmatrix} \times \begin{pmatrix} {}^-2 & 3 \\ 6 & {}^-1 \\ 0 & 2 \end{pmatrix}$$

$\begin{pmatrix} 2(-2) + 3(6) + 0(0) \end{pmatrix}$ Multiply 1st row of A by 1st column of B.

$\begin{pmatrix} 14 & 2(3) + 3({}^-1) + 0(2) \end{pmatrix}$ Multiply 1st row of A by 2nd column of B.

$\begin{pmatrix} 14 & 3 \\ 1({}^-2) - 4(6) - 2(0) & \end{pmatrix}$ Multiply 2nd row of A by 1st column of B.

$\begin{pmatrix} 14 & 3 \\ {}^-26 & 1(3) - 4({}^-1) - 2(2) \end{pmatrix}$ Multiply 2nd row of A by 2nd column of B.

$\begin{pmatrix} 14 & 3 \\ {}^-26 & 3 \\ 0({}^-2) + 1(6) + 1(0) & \end{pmatrix}$ Multiply 3rd row of A by 1st column of B.

$$\begin{pmatrix} 14 & 3 \\ -26 & 3 \\ 6 & 0(3)+1(-1)+1(2) \end{pmatrix}$$ Multiply 3rd row of A by 2nd column of B.

$$\begin{pmatrix} 14 & 3 \\ -26 & 3 \\ 6 & 1 \end{pmatrix}$$

The product of BA is not defined since the number of columns in B is not equal to the number of rows in A.

Practice problems:

1. $\begin{pmatrix} 3 & 4 \\ -2 & 1 \end{pmatrix} \begin{pmatrix} -1 & 7 \\ -3 & 1 \end{pmatrix}$

2. $\begin{pmatrix} 1 & -2 \\ 3 & 4 \\ 2 & 5 \\ -1 & 6 \end{pmatrix} \begin{pmatrix} 3 & -1 & -4 \\ -1 & 2 & 3 \end{pmatrix}$

SKILL 8.6 Rewrite a matrix equation as an equivalent system of linear equations or vice versa.

When given the following system of equations:

$$ax + by = e$$
$$cx + dy = f$$

the matrix equation is written in the form:

$$\begin{pmatrix} a & b \\ c & d \end{pmatrix} \begin{pmatrix} x \\ y \end{pmatrix} = \begin{pmatrix} e \\ f \end{pmatrix}$$

The solution is found using the inverse of the matrix of coefficients. Inverse of matrices can be written as follows:

$$A^{-1} = \frac{1}{\text{determinant of } A} \begin{pmatrix} d & -b \\ -c & a \end{pmatrix}$$

Sample Problem:

1. Write the matrix equation of the system.
$$3x - 4y = 2$$
$$2x + y = 5$$

$$\begin{pmatrix} 3 & ^-4 \\ 2 & 1 \end{pmatrix} \begin{pmatrix} x \\ y \end{pmatrix} = \begin{pmatrix} 2 \\ 5 \end{pmatrix}$$ Definition of matrix equation.

$$\begin{pmatrix} x \\ y \end{pmatrix} = \frac{1}{11} \begin{pmatrix} 1 & 4 \\ ^-2 & 3 \end{pmatrix} \begin{pmatrix} 2 \\ 5 \end{pmatrix}$$ Multiply by the inverse of the coefficient matrix.

$$\begin{pmatrix} x \\ y \end{pmatrix} = \frac{1}{11} \begin{pmatrix} 22 \\ 11 \end{pmatrix}$$ Matrix multiplication.

$$\begin{pmatrix} x \\ y \end{pmatrix} = \begin{pmatrix} 2 \\ 1 \end{pmatrix}$$ Scalar multiplication.

The solution is (2,1).

Practice problems:

1. $x + 2y = 5$
   $3x + 5y = 14$

2. $^-3x + 4y - z = 3$
   $x + 2y - 3z = 9$
   $y - 5z = ^-1$

SKILL 8.7 Represent problem situations using discrete structures such as sequences, finite graphs, and matrices.

Sequences can be **finite** or **infinite**. A finite sequence is a sequence whose domain consists of the set {1, 2, 3, ... *n*} or the first *n* positive integers. An infinite sequence is a sequence whose domain consists of the set {1, 2, 3, ...}; which is in other words all positive integers.

A **recurrence relation** is an equation that defines a sequence recursively; in other words, each term of the sequence is defined as a function of the preceding terms.

A real-life application would be using a recurrence relation to determine how much your savings would be in an account at the end of a certain period of time. For example:

You deposit $5,000 in your savings account. Your bank pays 5% interest compounded annually. How much will your account be worth at the end of 10 years?

Let V represent the amount of money in the account and V_n represent the amount of money after n years.

The amount in the account after n years equals the amount in the account after $n - 1$ years plus the interest for the nth year. This can be expressed as the recurrence relation V_0 where your initial deposit is represented by $V_0 = 5,000$.

$$V_0 = V_0$$
$$V_1 = 1.05 V_0$$
$$V_2 = 1.05 V_1 = (1.05)^2 V_0$$
$$V_3 = 1.05 V_2 = (1.05)^3 V_0$$
$$......$$
$$V_n = (1.05) V_{n-1} = (1.05)^n V_0$$

Inserting the values into the equation, you get
$V_{10} = (1.05)^{10} (5,000) = 8,144$.

You determine that after investing $5,000 in an account earning 5% interest, compounded annually for 10 years, you would have $8,144.

Matrices are used often to solve systems of equations. They are also used by physicists, mathematicians, and biologists to organize and study data such as population growth. It is also used in finance for such purposes as investment growth analysis and portfolio analysis. Matrices are easily translated into computer code in high-level programming languages and can be easily expressed in electronic spreadsheets.

A simple financial example of using a matrix to solve a problem follows:

A company has two stores. The income and expenses (in dollars) for the two stores, for three months, are shown in the matrices.

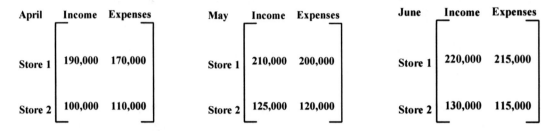

The owner wants to know what his first-quarter income and expenses were, so he adds the three matrices.

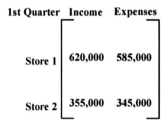

Then, to find the profit for each store:

Profit for Store 1 = $620,000 - $585,000 = $35,000
Profit for Store 2 = $355,000 - $345,000 = $10,000

COMPETENCY 9.0 KNOWLEDGE OF CALCULUS

SKILL 9.1 Solve problems using the limit theorems concerning sums, products, and quotients of functions.

The limit of a function is the y value that the graph approaches as the x values approach a certain number. To find a limit there are two points to remember.

1. Factor the expression completely and cancel all common factors in fractions.
2. Substitute the number to which the variable is approaching. In most cases this produces the value of the limit.

If the variable in the limit is approaching ∞, factor and simplify first; then examine the result. If the result does not involve a fraction with the variable in the denominator, the limit is usually also equal to ∞. If the variable is in the denominator of the fraction, the denominator is getting larger which makes the entire fraction smaller. In other words the limit is zero.

Examples:

1. $\lim\limits_{x \to ^-3} \dfrac{x^2+5x+6}{x+3} + 4x$ Factor the numerator.

 $\lim\limits_{x \to ^-3} \dfrac{(x+3)(x+2)}{(x+3)} + 4x$ Cancel the common factors.

 $\lim\limits_{x \to ^-3} (x+2) + 4x$ Substitute $^-3$ for x.

 $(^-3+2) + 4(^-3)$ Simplify.

 $^-1 + ^-12$

 $^-13$

2. $\lim\limits_{x \to \infty} \dfrac{2x^2}{x^5}$ Cancel the common factors.

 $\lim\limits_{x \to \infty} \dfrac{2}{x^3}$

 Since the denominator is getting larger, the entire fraction is getting smaller. The fraction is getting close to zero.

 $\dfrac{2}{\infty^3}$

MATHEMATICS 6-12

Practice problems:

1. $\lim\limits_{x \to \pi} 5x^2 + \sin x$

2. $\lim\limits_{x \to {-4}} \dfrac{x^2 + 9x + 20}{x + 4}$

SKILL 9.2 Find the derivatives of algebraic, trigonometric, exponential, and logarithmic functions.

Derivatives of algebraic functions.

A. Derivative of a constant--for any constant, the derivative is always zero.

B. Derivative of a variable--the derivative of a variable (i.e. x) is one.

C. Derivative of a variable raised to a power--for variable expressions with rational exponents (i.e. $3x^2$) multiply the coefficient (3) by the exponent (2) then subtract one (1) from the exponent to arrive at the derivative $(3x^2) = (6x)$

Example:

1. $y = 5x^4$ Take the derivative.

   $\dfrac{dy}{dx} = (5)(4)x^{4-1}$ Multiply the coefficient by the exponent and subtract 1 from the exponent.

   $\dfrac{dy}{dx} = 20x^3$ Simplify.

2. $y = \dfrac{1}{4x^3}$ Rewrite using negative exponent.

   $y = \dfrac{1}{4} x^{-3}$ Take the derivative.

   $\dfrac{dy}{dx} = \left(\dfrac{1}{4} \cdot {-3}\right) x^{-3-1}$

   $\dfrac{dy}{dx} = \dfrac{-3}{4} x^{-4} = \dfrac{-3}{4x^4}$ Simplify.

3. $y = 3\sqrt{x^5}$ Rewrite using $\sqrt[z]{x^n} = x^{n/z}$.

   $y = 3x^{5/2}$ Take the derivative.

   $\dfrac{dy}{dx} = (3)\left(\dfrac{5}{2}\right)x^{5/2-1}$

   $\dfrac{dy}{dx} = \left(\dfrac{15}{2}\right)x^{3/2}$ Simplify.

   $\dfrac{dy}{dx} = 7.5\sqrt{x^3} = 7.5x\sqrt{x}$

Derivatives of trigonometric functions.

A. sin x --the derivative of the sine of x is simply the cosine of x.

B. cos x --the derivative of the cosine of x is negative one ($^-1$) times the sine of x.

C. tan x --the derivative of the tangent of x is the square of the secant of x.

If the object of the trig. function is an expression other than x, follow the above rules substituting the expression for x. The only additional step is to multiply the result by the derivative of the expression.

Examples:

1. $y = \pi \sin x$ Carry the coefficient (π) throughout the problem.

   $\dfrac{dy}{dx} = \pi \cos x$

2. $y = \dfrac{2}{3}\cos x$

   Do not forget to multiply the coefficient by negative one when taking the derivative of a cosine function.

   $\dfrac{dy}{dx} = \dfrac{^-2}{3}\sin x$

3. $y = 4\tan(5x^3)$

$\dfrac{dy}{dx} = 4\sec^2(5x^3)(5 \cdot 3x^{3-1})$ The derivative of $\tan x$ is $\sec^2 x$.

$\dfrac{dy}{dx} = 4\sec^2(5x^3)(15x^2)$ Carry the $(5x^3)$ term throughout the problem.

$\dfrac{dy}{dx} = 4 \cdot 15x^2 \sec^2(5x^3)$ Multiply $4\sec^2(5x^3)$ by the derivative of $5x^3$.

$\dfrac{dy}{dx} = 60x^2 \sec^2(5x^3)$ Rearrange the terms and simplify.

Derivatives of exponential functions.

$f(x) = e^x$ is an exponential function. The derivative of e^x is exactly the same thing → e^x. If instead of x, the exponent on e is an expression, the derivative is the same e raised to the algebraic exponent multiplied by the derivative of the algebraic expression.

If a base other than e is used, the derivative is the natural log of the base times the original exponential function times the derivative of the exponent.

Examples:

1. $y = e^x$

    $\dfrac{dy}{dx} = e^x$

2. $y = e^{3x}$

    $\dfrac{dy}{dx} = e^{3x} \cdot 3 = 3e^{3x}$ Multiply e^{3x} by the derivative of $3x$ which is 3.

    $\dfrac{dy}{dx} = 3e^{3x}$ Rearrange the terms.

3. $y = \dfrac{5}{e^{\sin x}}$

$y = 5e^{-\sin x}$ Rewrite using negative exponents

$\dfrac{dy}{dx} = 5e^{-\sin x} \bullet \left(^-\cos x\right)$ Multiply $5e^{-\sin x}$ by the derivative of $^-\sin x$ which is $^-\cos x$.

$\dfrac{dy}{dx} = \dfrac{^-5\cos x}{e^{\sin x}}$ Use the definition of negative exponents to simplify.

4. $y = {}^-2 \bullet \ln 3^{4x}$

$\dfrac{dy}{dx} = {}^-2 \bullet (\ln 3)(3^{4x})(4)$ The natural log of the base is ln3. The derivative of $4x$ is 4.

$\dfrac{dy}{dx} = {}^-8 \bullet 3^{4x} \ln 3$ Rearrange terms to simplify.

Derivatives of logarithmic functions.

The most common logarithmic function on the Exam is the natural logarithmic function ($\ln x$). The derivative of $\ln x$ is simply $1/x$. If x is replaced by an algebraic expression, the derivative is the fraction one divided by the expression multiplied by the derivative of the expression.

For all other logarithmic functions, the derivative is 1 over the argument of the logarithm multiplied by 1 over the natural logarithm (ln) of the base multiplied by the derivative of the argument. Examples:

1. $y = \ln x$

$\dfrac{dy}{dx} = \dfrac{1}{x}$

2. $y = 3\ln(x^{-2})$

$\dfrac{dy}{dx} = 3 \cdot \dfrac{1}{x^{-2}} \cdot \left(-2x^{-2-1}\right)$

Multiply one over the argument (x^{-2}) by the derivative of x^{-2} which is $-2x^{-2-1}$.

$\dfrac{dy}{dx} = 3 \cdot x^2 \cdot \left(-2x^{-3}\right)$

$\dfrac{dy}{dx} = \dfrac{-6x^2}{x^3}$

Simplify using the definition of negative exponents.

$\dfrac{dy}{dx} = \dfrac{-6}{x}$

Cancel common factors to simplify.

3. $y = \log_5(\tan x)$

$\dfrac{dy}{dx} = \dfrac{1}{\tan x} \cdot \dfrac{1}{\ln 5} \cdot (\sec^2 x)$

The derivative of $\tan x$ is $\sec^2 x$.

$\dfrac{dy}{dx} = \dfrac{\sec^2 x}{(\tan x)(\ln 5)}$

TEACHER CERTIFICATION STUDY GUIDE

SKILL 9.3 **Find the derivative of the sum, product, quotient, or the composition of functions.**

A. Derivative of a sum--find the derivative of each term separately and add the results.

B. Derivative of a product--multiply the derivative of the first factor by the second factor and add to it the product of the first factor and the derivative of the second factor.

Remember the phrase "first times the derivative of the second plus the second times the derivative of the first."

C. Derivative of a quotient--use the rule "bottom times the derivative of the top minus the top times the derivative of the bottom all divided by the bottom squared."

Examples:

1. $y = 3x^2 + 2\ln x + 5\sqrt{x}$ $\qquad \sqrt{x} = x^{1/2}$.

$$\frac{dy}{dx} = 6x^{2-1} + 2 \cdot \frac{1}{x} + 5 \cdot \frac{1}{2} x^{1/2-1}$$

$$\frac{dy}{dx} = 6x + \frac{2}{x} + \frac{5}{2} \cdot \frac{1}{\sqrt{x}} = \frac{12x^2 + 4 + 5\sqrt{x}}{2x}$$

$\qquad x^{1/2-1} = x^{-1/2} = 1/\sqrt{x}$.

$$= \frac{12x^2 + 5\sqrt{x} + 4}{2x}$$

2. $y = 4e^{x^2} \cdot \sin x$

$$\frac{dy}{dx} = 4(e^{x^2} \cdot \cos x + \sin x \cdot e^{x^2} \cdot 2x)$$

The derivative of e^{x^2} is $e^{x^2} \cdot 2$.

$$\frac{dy}{dx} = 4(e^{x^2} \cos x + 2xe^{x^2} \sin x)$$

$$\frac{dy}{dx} = 4e^{x^2} \cos x + 8xe^{x^2} \sin x$$

3. $y = \dfrac{\cos x}{x}$

$$\frac{dy}{dx} = \frac{x(-\sin x) - \cos x \cdot 1}{x^2}$$

The derivative of x is 1.
The derivative of $\cos x$ is $-\sin x$.

$$\frac{dy}{dx} = \frac{-x \sin x - \cos x}{x^2}$$

D. Derivatives of the composition of functions (chain rule)-- a composite function is made up of two or more separate functions such as $\sin(\ln x)$ or $x^2 e^{3x}$. To find the derivatives of these composite functions requires two steps. First identify the predominant function in the problem. For example, in $\sin(\ln x)$) the major function is the sine function. In $x^2 e^{3x}$ the major function is a product of two expressions (x^2 and e^{3x}). Once the predominant function is identified, apply the appropriate differentiation rule. Be certain to include the step of taking the derivative of every part of the functions which comprise the composite function. Use parentheses as much as possible.

Examples:

1. $y = \sin(\ln x)$ The major function is a sine function.

$\dfrac{dy}{dx} = [\cos(\ln x)] \cdot \left[\dfrac{1}{x}\right]$ The derivative of $\sin x$ is $\cos x$.

The derivative of $\ln x$ is $1/x$.

2. $y = x^2 \cdot e^{3x}$ The major function is a product.

$\dfrac{dy}{dx} = x^2 \left(e^{3x} \cdot 3\right) + e^{3x} \cdot 2x$

The derivative of a product is "First times the derivative of second plus the second times the derivative of the first."

$\dfrac{dy}{dx} = 3x^2 e^{3x} + 2x e^{3x}$

3. $y = \tan^2\left(\dfrac{\ln x}{\cos x}\right)$

This function is made of several functions. The major function is a power function.

$$\dfrac{dy}{dx} = \left[2\tan^{2-1}\left(\dfrac{\ln x}{\cos x}\right)\right]\left[\sec^2\left(\dfrac{\ln x}{\cos x}\right)\right]\left[\dfrac{d}{dx}\left(\dfrac{\ln x}{\cos x}\right)\right]$$

The derivative of $\tan x$ is $\sec^2 x$. Hold off one more to take the derivative of $\ln x / \cos x$.

$$\dfrac{dy}{dx} = \left[2\tan\left(\dfrac{\ln x}{\cos x}\right)\sec^2\left(\dfrac{\ln x}{\cos x}\right)\right]\left[\dfrac{(\cos x)(1/x) - \ln x(^-\sin x)}{\cos^2 x}\right]$$

$$\dfrac{dy}{dx} = \left[2\tan\left(\dfrac{\ln x}{\cos x}\right)\sec^2\left(\dfrac{\ln x}{\cos x}\right)\right]\left[\dfrac{(\cos x)(1/x) + \ln x(\sin x)}{\cos^2 x}\right]$$

The derivative of a quotient is "Bottom times the derivative of the top minus the top times the derivative of the bottom all divided by the bottom squared."

TEACHER CERTIFICATION STUDY GUIDE

SKILL 9.4   Identify and apply definitions of the derivative of a function.

The derivative of a function has two basic interpretations.

I.   Instantaneous rate of change
II. Slope of a tangent line at a given point

If a question asks for the rate of change of a function, take the derivative to find the equation for the rate of change. Then plug in for the variable to find the instantaneous rate of change.

The following is a list summarizing some of the more common quantities referred to in rate of change problems.

area	height	profit
decay	population growth	sales
distance	position	temperature
frequency	pressure	volume

Pick a point, say $x = {}^-3$, on the graph of a function. Draw a tangent line at that point. Find the derivative of the function and plug in $x = {}^-3$. The result will be the slope of the tangent line.

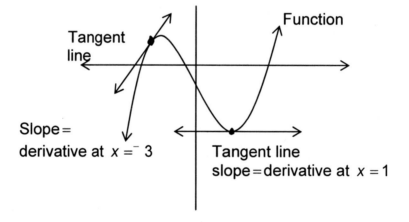

SKILL 9.5   Use the derivative to find the slope of a curve at a point.

To find the slope of a curve at a point, there are two steps to follow.

1. Take the derivative of the function.
2. Plug in the value to find the slope.

If plugging into the derivative yields a value of zero, the tangent line is horizontal at that point.

If plugging into the derivative produces a fraction with zero in the denominator, the tangent line at this point has an undefined slope and is thus a vertical line.

Examples:

1. Find the slope of the tangent line for the given function at the given point.

$$y = \frac{1}{x-2} \text{ at } (3,1)$$

$y = (x-2)^{-1}$ Rewrite using negative exponents.

$\dfrac{dy}{dx} = {}^-1(x-2)^{-1-1}(1)$ Use the Chain rule. The derivative of $(x-2)$ is 1.

$\dfrac{dy}{dx} = {}^-1(x-2)^{-2}$

$\dfrac{dy}{dx}\bigg|_{x=3} = {}^-1(3-2)^{-2}$ Evaluate at $x = 3$.

$\dfrac{dy}{dx}\bigg|_{x=3} = {}^-1$

The slope of the tangent line is $^-1$ at $x = 3$.

2. Find the points where the tangent to the curve $f(x) = 2x^2 + 3x$ is parallel to the line $y = 11x - 5$.

$f'(x) = 2 \bullet 2x^{2-1} + 3$ Take the derivative of $f(x)$ to get the slope of a tangent line.

$f'(x) = 4x + 3$

$4x + 3 = 11$ Set the slope expression $(4x + 3)$ equal to the slope of $y = 11x - 5$.

$x = 2$ Solve for the x value of the point.

$f(2) = 2(2)^2 + 3(2)$ The y value is 14.

$f(2) = 14$ So $(2, 14)$ is the point on $f(x)$ where the tangent line is parallel to $y = 11x - 5$.

SKILL 9.6   **Find the equation of a tangent line or a normal line at a point on a curve.**

To write an equation of **a tangent line** at a point, two things are needed.

A point--the problem will usually provide a point, (x, y). If the problem only gives an x value, plug the value into the original function to get the y coordinate.

The slope--to find the slope, take the derivative of the original function. Then plug in the x value of the point to get the slope.

After obtaining a point and a slope, use the Point-Slope form for the equation of a line:

$$(y - y_1) = m(x - x_1)$$

where m is the slope and (x_1, y_1) is the point.

Example:
Find the equation of the tangent line to $f(x) = 2e^{x^2}$ at $x = {}^-1$.

$f({}^-1) = 2e^{({}^-1)^2}$	Plug in the x coordinate to obtain the y coordinate.
$= 2e^1$	The point is $({}^-1, 2e)$.
$f'(x) = 2e^{x^2} \bullet (2x)$	
$f'({}^-1) = 2e^{({}^-1)^2} \bullet (2 \bullet {}^-1)$	
$f'({}^-1) = 2e^1({}^-2)$	
$f'({}^-1) = {}^-4e$	The slope at $x = {}^-1$ is ${}^-4e$.
$(y - 2e) = {}^-4e(x - {}^-1)$	Plug in the point $({}^-1, 2e)$ and the slope $m = {}^-4e$. Use the point slope form of a line.
$y = {}^-4ex - 4e + 2e$	
$y = {}^-4ex - 2e$	Simplify to obtain the equation for the tangent line.

A **normal line** is a line which is perpendicular to a tangent line at a given point. Perpendicular lines have slopes which are negative reciprocals of each other. To find the equation of a normal line, first get the slope of the tangent line at the point. Find the negative reciprocal of this slope. Next, use the new slope and the point on the curve, both the x_1 and y_1 coordinates, and substitute into the Point-Slope form of the equation for a line:
$$(y - y_1) = slope \bullet (x - x_1)$$

Examples:
1. Find the equation of the normal line to the tangent to the curve $y = (x^2 - 1)(x - 3)$ at $x = {}^-2$.

$f(-2) = (({}^-2)^2 - 1)({}^-2 - 3)$	First find the y coordinate of the point on the curve. Here,
$f(-2) = {}^-15$	$y = {}^-15$ when $x = {}^-2$.
$y = x^3 - 3x^2 - x + 3$	Before taking the derivative, multiply the expression first. The derivative of a sum is easier to find than the derivative of a product.

Mathematics 6-12

$y' = 3x^2 - 6x - 1$ Take the derivative to find the slope of the tangent line.

$y'_{x=^-2} = 3(^-2)^2 - 6(^-2) - 1$

$y'_{x=^-2} = 23$

slope of normal $= \dfrac{^-1}{23}$

For the slope of the normal line, take the negative reciprocal of the tangent line's slope.

$(y - {^-15}) = \dfrac{^-1}{23}(x - {^-2})$ Plug (x_1, y_1) into the point-slope equation.

$(y + 15) = \dfrac{^-1}{23}(x + 2)$

$y = -\dfrac{1}{23}x - 14\dfrac{21}{23}$

$y = -\dfrac{1}{23}x + \dfrac{2}{23} - 15 = \dfrac{1}{23}x - 14\dfrac{21}{23}$

2. Find the equation of the normal line to the tangent to the curve $y = \ln(\sin x)$ at $x = \pi$.

$f(\pi) = \ln(\sin \pi)$ $\sin \pi = 1$ and $\ln(1) = 0$ (recall $e^0 = 1$).

$f(\pi) = \ln(1) = 0$ So $x_1 = \pi$ and $y_1 = 0$.

$y' = \dfrac{1}{\sin x} \cdot \cos x$

Take the derivative to find the slope of the tangent line.

$y'_{x=\pi} = \dfrac{\cos \pi}{\sin \pi} = \dfrac{0}{1}$

Slope of normal does not exist. $\dfrac{^-1}{0}$ does not exist. So the normal line is vertical at $x = \pi$.

SKILL 9.7 Determine if a function is increasing or decreasing by using the first derivative in a given interval.

A function is said to be increasing if it is rising from left to right and decreasing if it is falling from left to right. Lines with positive slopes are increasing, and lines with negative slopes are decreasing. If the function in question is something other than a line, simply refer to the slopes of the tangent lines as the test for increasing or decreasing. Take the derivative of the function and plug in an x value to get the slope of the tangent line; a positive slope means the function is increasing and a negative slope means it is decreasing. If an interval for x values is given, just pick any point between the two values to substitute.

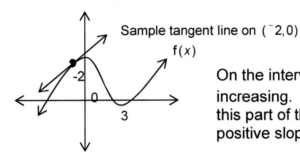

Sample tangent line on $(^-2, 0)$

On the interval $(^-2, 0)$, $f(x)$ is increasing. The tangent lines on this part of the graph have positive slopes.

Example:

The growth of a certain bacteria is given by $f(x) = x + \dfrac{1}{x}$. Determine if the rate of growth is increasing or decreasing on the time interval $(^-1, 0)$.

$$f'(x) = 1 + \dfrac{^-1}{x^2}$$

To test for increasing or decreasing, find the slope of the tangent line by taking the derivative.

$$f'\left(\dfrac{^-1}{2}\right) = 1 + \dfrac{^-1}{(^-1/2)^2}$$

Pick any point on $(^-1, 0)$ and substitute into the derivative.

$$f'\left(\dfrac{^-1}{2}\right) = 1 + \dfrac{^-1}{1/4}$$
$$= 1 - 4$$
$$= ^-3$$

The slope of the tangent line at $x = \dfrac{^-1}{2}$ is $^-3$. The exact value of the slope is not important. The important fact is that the slope is negative.

Mathematics 6–12

SKILL 9.8 Find relative and absolute maxima and minima.

Substituting an x value into a function produces a corresponding y value. The coordinates of the point (x,y), where y is the largest of all the y values, is said to be a maximum point. The coordinates of the point (x,y), where y is the smallest of all the y values, is said to be a minimum point. To find these points, only a few x values must be tested. First, find all of the x values that make the derivative either zero or undefined. Substitute these values into the original function to obtain the corresponding y values. Compare the y values. The largest y value is a maximum; the smallest y value is a minimum. If the question asks for the maxima or minima on an interval, be certain to also find the y values that correspond to the numbers at either end of the interval.

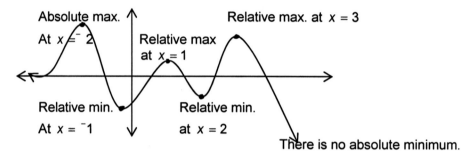

Example:

Find the maxima and minima of $f(x) = 2x^4 - 4x^2$ at the interval $(^-2, 1)$.

$f'(x) = 8x^3 - 8x$	Take the derivative first. Find all the x values (critical values) that make the derivative zero or undefined. In this case, there are no x values that make the derivative undefined.
$8x^3 - 8x = 0$	
$8x(x^2 - 1) = 0$	
$8x(x-1)(x+1) = 0$	Substitute the critical values into the original function. Also, plug in the endpoint of the interval. Note that 1 is a critical point and an endpoint.
$x = 0, x = 1, \text{ or } x = ^-1$	
$f(0) = 2(0)^4 - 4(0)^2 = 0$	
$f(1) = 2(1)^4 - 4(1)^2 = ^-2$	
$f(^-1) = 2(^-1)^4 - 4(^-1)^2 = ^-2$	
$f(^-2) = 2(^-2)^4 - 4(^-2)^2 = 16$	

The maximum is at $(-2, 16)$ and there are minima at $(1, -2)$ and $(-1, -2)$. $(0,0)$ is neither the maximum or minimum on $(-2, 1)$ but it is still considered a relative extra point.

SKILL 9.9 Find intervals on a curve where the curve is concave up or concave down.

The first derivative reveals whether a curve is rising or falling (increasing or decreasing) from the left to the right. In much the same way, the second derivative relates whether the curve is concave up or concave down. Curves which are concave up are said to "collect water;" curves which are concave down are said to "dump water." To find the intervals where a curve is concave up or concave down, follow the following steps.

1. Take the second derivative (i.e. the derivative of the first derivative).
2. Find the critical x values.
   - Set the second derivative equal to zero and solve for critical x values.
   - Find the x values that make the second derivative undefined (i.e. make the denominator of the second derivative equal to zero).
   Such values may not always exist.
3. Pick sample values which are both less than and greater than each of the critical values.
4. Substitute each of these sample values into the second derivative and determine whether the result is positive or negative.
   - If the sample value yields a positive number for the second derivative, the curve is concave up on the interval where the sample value originated.
   - If the sample value yields a negative number for the second derivative, the curve is concave down on the interval where the sample value originated.

Example:

Find the intervals where the curve is concave up and concave down for $f(x) = x^4 - 4x^3 + 16x - 16$.

$f'(x) = 4x^3 - 12x^2 + 16$ Take the second derivative.
$f''(x) = 12x^2 - 24x$

                                       Find the critical values by setting
                                       the second derivative equal to
$12x^2 - 24x = 0$ zero.
$12x(x - 2) = 0$ There are no values that make
$x = 0$ or $x = 2$ the second derivative undefined.
                                       Set up a number line with the
                                       critical values.

Sample values: $^-1, 1, 3$ Pick sample values in each of the
                                       3 intervals. If the sample value
$f''(^-1) = 12(^-1)^2 - 24(^-1) = 36$ produces a negative number,
$f''(1) = 12(1)^2 - 24(1) = ^-12$ the function is concave down.
$f''(3) = 12(3)^2 - 24(3) = 36$

If the value produces a positive number, the curve is concave up. If the value produces a zero, the function is linear.

Therefore when $x < 0$ the function is concave up,
when $0 < x < 2$ the function is concave down,
when $x > 2$ the function is concave up.

SKILL 9.10 Identify points of inflection.

A point of inflection is a point where a curve changes from being concave up to concave down or vice versa. To find these points, follow the steps for finding the intervals where a curve is concave up or concave down. A critical value is part of an inflection point if the curve is concave up on one side of the value and concave down on the other. The critical value is the x coordinate of the inflection point. To get the y coordinate, plug the critical value into the original function.

Example: Find the inflection points of $f(x) = 2x - \tan x$ where $\dfrac{-\pi}{2} < x < \dfrac{\pi}{2}$.

$(x) = 2x - \tan x$ $\quad \dfrac{-\pi}{2} < x < \dfrac{\pi}{2}$ $\qquad$ Note the restriction on x.

$f'(x) = 2 - \sec^2 x$ $\qquad$ Take the second derivative.

$f''(x) = 0 - 2 \bullet \sec x \bullet (\sec x \tan x)$ $\qquad$ Use the Power rule.

$= {}^-2 \bullet \dfrac{1}{\cos x} \bullet \dfrac{1}{\cos x} \bullet \dfrac{\sin x}{\cos x}$

The derivative of $\sec x$ is $(\sec x \tan x)$.

$f''(x) = \dfrac{{}^-2\sin x}{\cos^3 x}$

Find critical values by solving for the second derivative equal to zero.

$0 = \dfrac{{}^-2\sin x}{\cos^3 x}$

No x values on $\left(\dfrac{-\pi}{2}, \dfrac{\pi}{2}\right)$ make the denominator zero.

${}^-2\sin x = 0$
$\sin x = 0$
$x = 0$

Pick sample values on each side of the critical value $x = 0$.

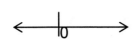

Sample values: $x = \dfrac{-\pi}{4}$ and $x = \dfrac{\pi}{4}$

$f''\left(\dfrac{-\pi}{4}\right) = \dfrac{{}^-2\sin({}^-\pi/4)}{\cos^3(\pi/4)} = \dfrac{{}^-2({}^-\sqrt{2}/2)}{(\sqrt{2}/2)^3} = \dfrac{\sqrt{2}}{(\sqrt{8}/8)} = \dfrac{8\sqrt{2}}{\sqrt{8}} = \dfrac{8\sqrt{2}}{\sqrt{8}} \bullet \dfrac{\sqrt{8}}{\sqrt{8}}$

$= \dfrac{8\sqrt{16}}{8} = 4$

$f''\left(\dfrac{\pi}{4}\right) = \dfrac{{}^-2\sin(\pi/4)}{\cos^3(\pi/4)} = \dfrac{{}^-2(\sqrt{2}/2)}{(\sqrt{2}/2)^3} = \dfrac{{}^-\sqrt{2}}{(\sqrt{8}/8)} = \dfrac{{}^-8\sqrt{2}}{\sqrt{8}} = -4$

The second derivative is positive on $(0, \infty)$ and negative on $({}^-\infty, 0)$. So the curve changes concavity at $x = 0$. Use the original equation to find the y value that inflection occurs at.

$f(0) = 2(0) - \tan 0 = 0 - 0 = 0$ $\qquad$ The inflection point is $(0,0)$.

SKILL 9.11 Solve problems using velocity and acceleration of a particle moving along a line.

If a particle (or a car, a bullet, etc.) is moving along a line, then the distance that the particle travels can be expressed by a function in terms of time.

1. The first derivative of the distance function will provide the velocity function for the particle. Substituting a value for time into this expression will provide the instantaneous velocity of the particle at the time. Velocity is the rate of change of the distance traveled by the particle.
Taking the absolute value of the derivative provides the speed of the particle. A positive value for the velocity indicates that the particle is moving forward, and a negative value indicates the particle is moving backwards.

2. The second derivative of the distance function (which would also be the first derivative of the velocity function) provides the acceleration function. The acceleration of the particle is the rate of change of the velocity. If a value for time produces a positive acceleration, the particle is speeding up; if it produces a negative value, the particle is slowing down. If the acceleration is zero, the particle is moving at a constant speed. To find the time when a particle stops, set the first derivative (i.e. the velocity function) equal to zero and solve for time. This time value is also the instant when the particle changes direction.

Example:

The motion of a particle moving along a line is according to the equation: $s(t) = 20 + 3t - 5t^2$ where s is in meters and t is in seconds. Find the position, velocity, and acceleration of a particle at $t = 2$ seconds.

$s(2) = 20 + 3(2) - 5(2)^2$ $ = 6$ meters	Plug $t = 2$ into the original equation to find the position.
$s\,'(t) = v(t) = 3 - 10t$	The derivative of the first function gives the velocity.
$v(2) = 3 - 10(2) = {}^-17$ m/s	Plug $t = 2$ into the velocity function to find the velocity. $^-17$ m/s indicates the particle is moving backwards.
$s\,''(t) = a(t) = {}^-10$ $a(2) = {}^-10$ m/s²	The second derivation of position gives the acceleration. Substitute $t = 2$, yields an acceleration of $^-10$ m/s², which indicates the particle is slowing down.

Practice problem:

A particle moves along a line with acceleration $a(t) = 5t + 2$. The velocity after 2 seconds is $^-10$ m/sec.

1. Find the initial velocity.
2. Find the velocity at $t = 4$.

TEACHER CERTIFICATION STUDY GUIDE

SKILL 9.12 Solve problems using instantaneous rates of change and related rates of change, such as growth and decay.

Finding the rate of change of one quantity (for example distance, volume, etc.) with respect to time it is often referred to as a rate of change problem. To find an instantaneous rate of change of a particular quantity, write a function in terms of time for that quantity; then take the derivative of the function. Substitute in the values at which the instantaneous rate of change is sought.

Functions which are in terms of more than one variable may be used to find related rates of change. These functions are often not written in terms of time. To find a related rate of change, follow these steps.

1. Write an equation which relates all the quantities referred to in The problem.

2. Take the derivative of both sides of the equation with respect to time. Follow the same steps as used in implicit differentiation. This means take the derivative of each part of the equation remembering to multiply each term by the derivative of the variable involved with respect to time. For example, if a term includes the variable v for volume, take the derivative of the term remembering to multiply by dv/dt for the derivative of volume with respect to time. dv/dt is the rate of change of the volume.

3. Substitute the known rates of change and quantities, and solve for the desired rate of change.

Example:

1. What is the instantaneous rate of change of the area of a circle where the radius is 3 cm?

$A(r) = \pi r^2$ Write an equation for area.
$A'(r) = 2\pi r$ Take the derivative to find the rate of change.
$A'(3) = 2\pi(3) = 6\pi$ Substitute in $r = 3$ to arrive at the instantaneous rate of change.

SKILL 9.13 Find antiderivatives for algebraic, trigonometric, exponential, and logarithmic functions.

Antiderivatives for algebraic functions.

Taking the antiderivative of a function is the opposite of taking the derivative of the function--much in the same way that squaring an expression is the opposite of taking the square root of the expression. For example, since the derivative of x^2 is $2x$ then the antiderivative of $2x$ is x^2. The key to being able to take antiderivatives is being as familiar as possible with the derivative rules.

To take the antiderivative of an algebraic function (the sum of products of coefficients and variables raised to powers other than negative one), take the antiderivative of each term in the function by following these steps.

1. Take the antiderivative of each term separately.

2. The coefficient of the variable should be equal to one plus the exponent.
3. If the coefficient is not one more than the exponent, put the correct coefficient on the variable and also multiply by the reciprocal of the number put in.
    Ex. For $4x^5$, the coefficient should be 6 not 4. So put in 6 and the reciprocal 1/6 to achieve $(4/6)6x^5$.
4. Finally take the antiderivative by replacing the coefficient and variable with just the variable raised to one plus the original exponent.
    Ex. For $(4/6)6x^5$, the antiderivative is $(4/6)x^6$.
    You have to add in constant c because there is no way to know if a constant was originally present since the derivative of a constant is zero.
5. Check your work by taking the first derivative of your answer. You should get the original algebraic function.

Examples: Take the antiderivative of each function.

1. $f(x) = 5x^4 + 2x$ The coefficient of each term is already 1 more than the exponent.
$F(x) = x^5 + x^2 + c$ $F(x)$ is the antiderivative of $f(x)$
$F'(x) = 5x^4 + 2x$ Check by taking the derivative of $F(x)$.

2. $f(x) = {}^-2x^{-3}$

$F(x) = x^{-2} + c = \dfrac{1}{x^2} + c$ $F(x)$ is the antiderivative of $f(x)$.

$F'(x) = {}^-2x^{-3}$ Check.

3. $f(x) = {}^-4x^2 + 2x^7$ Neither coefficient is correct.

$f(x) = {}^-4 \bullet \dfrac{1}{3} \bullet 3x^2 + 2 \bullet \dfrac{1}{8} \bullet 8x^7$

     Put in the correct coefficient along with its reciprocal.

$F(x) = {}^-4 \bullet \dfrac{1}{3}x^3 + 2 \bullet \dfrac{1}{8}x^8 + c$ $F(x)$ is the antiderivative of $f(x)$.

$F(x) = \dfrac{{}^-4}{3}x^3 + \dfrac{1}{4}x^8 + c$

$F'(x) = {}^-4x^2 + 2x^7$ Check.

Antiderivatives for trigonometric functions.

The rules for taking antiderivatives of trigonometric functions follow, but be aware that these can get very confusing to memorize because they are very similar to the derivative rules. Check the antiderivative you get by taking the derivative and comparing it to the original function.

1. $\sin x$ the antiderivative for $\sin x$ is $^-\cos x + c$.
2. $\cos x$ the antiderivative for $\cos x$ is $\sin x + c$.
3. $\tan x$ the antiderivative for $\tan x$ is $-\ln|\cos x| + c$.
4. $\sec^2 x$ the antiderivative for $\sec^2 x$ is $\tan x + c$.
5. $\sec x \tan x$ the antiderivative for $\sec x \tan x$ is $\sec x + c$.

If the trigonometric function has a coefficient, simply keep the coefficient and multiply the antiderivative by it.

Examples: Find the antiderivatives for the following functions.

1. $f(x) = 2\sin x$ Carry the 2 throughout the problem.

$F(x) = 2({}^-\cos x) = {}^-2\cos x + c$ $F(x)$ is the antiderivative.

$F'(x) = {}^-2({}^-\sin x) = 2\sin x$ Check by taking the derivative of $F(x)$.

2. $f(x) = \dfrac{\tan x}{5}$

$F(x) = \dfrac{-\ln|\cos x|}{5} + c$ $F(x)$ is the antiderivative of $f(x)$.

$F'(x) = \dfrac{1}{5}\left(-\dfrac{1}{|\cos x|}\right)(-\sin x) = \dfrac{1}{5}\tan x$ Check by taking the derivative of $F(x)$.

Practice problems: Find the antiderivative of each function.

1. $f(x) = {}^-20\cos x$ 2. $f(x) = \pi \sec x \tan x$

The antiderivative, or indefinite integral, of a given function f is a function F whose derivative is equal to f (i.e. $F' = f$).

Antiderivatives for logarithmic function.

The antiderivative of the natural logarithmic function ln x, derived using integration by parts, is x(ln x) − x + c.

$\int \ln x\, dx$

$u = \ln x$ and $dx = dv$ thus $\int \ln x\, dx = \int u\, dv$	Substitution.
$du = 1/x$ and $v = x$	Differentiation and Integration.
$uv - \int v\, du = (x)\ln x - \int x\dfrac{1}{x}dx$	Integration by parts.
$= (x)\ln x - x + c$	Solve by integration.

Mathematics 6-12   219

The antiderivative of composite functions containing natural logarithms is often found by substitution.

Example:

Find the indefinite integral (antiderivative) of $\int \frac{(\ln x)^2}{x} dx$.

Let u = ln x because the derivative of ln x, 1/x, is found in the expression. Thus du = (1/x) dx.

Substitution and integration yields...

$$\int \frac{(\ln x)^2}{x} dx = \int u^2 du = \frac{1}{3} u^3 = \frac{1}{3}(\ln x)^3$$

$\frac{1}{3}(\ln x)^3 + c$ is the antiderivative of $\int \frac{(\ln x)^2}{x} dx$.

Antiderivatives for exponential functions.

Use the following rules when finding the antiderivative of an exponential function.

1. e^x The antiderivative of e^x is the same $e^x + c$.
2. a^x The antiderivative of a^x, where a is any number, is $a^x/\ln a + c$.

Examples: Find the antiderivatives of the following functions:

1. $f(x) = 10e^x$

    $F(x) = 10e^x + c$ $F(x)$ is the antiderivative.

    $F'(x) = 10e^x$ Check by taking the derivative of $F(x)$.

2. $f(x) = \dfrac{2^x}{3}$

$F(x) = \dfrac{1}{3} \cdot \dfrac{2^x}{\ln 2} + c$ $F(x)$ is the antiderivative.

$F'(x) = \dfrac{1}{3\ln 2} \ln 2 (2^x)$

Check by taking the derivative of $F(x)$.

$F'(x) = \dfrac{2^x}{3}$

SKILL 9.14 Solve distance, area, and volume problems using integration.

To find the **distance** function, take the antiderivative of the velocity function. And to find the velocity function, find the antiderivative of the acceleration function. Use the information in the problem to solve for the constants that result from taking the antiderivatives.

Example:

A particle moves along the x axis with acceleration $a(t) = 6t - 6$. The initial velocity is 0 m/sec and the initial position is 8 cm to the right of the origin. Find the velocity and position functions.

$v(0) = 0$	Interpret the given information.
$s(0) = 8$	
$a(t) = 6t - 6$	Put in the coefficients needed to take the antiderivative.
$a(t) = 6 \bullet \dfrac{1}{2} \bullet 2t - 6$	
$v(t) = \dfrac{6}{2}t^2 - 6t + c$	Take the antiderivative of $a(t)$ to get $v(t)$.
$v(0) = 3(0)^3 - 6(0) + c = 0$	Use $v(0) = 0$ to solve for c.
$0 - 0 + c = 0$	$c = 0$
$c = 0$	
$v(t) = 3t^2 - 6t + 0$	Rewrite $v(t)$ using $c = 0$.
$v(t) = 3t^2 - 6\dfrac{1}{2} \bullet 2t$	Put in the coefficients needed to take the antiderivative.
$s(t) = t^3 - \dfrac{6}{2}t^2 + c$	Take the antiderivative of $v(t)$ to get $s(t) \rightarrow$ the distance function.
$s(0) = 0^3 - 3(0)^2 + c = 8$	Use $s(0) = 8$ to solve for c.
$c = 8$	
$s(t) = t^3 - 3t^2 + 8$	

Taking the integral of a function and evaluating it from one x value to another provides the total **area under the curve** (i.e. between the curve and the x axis). Remember, though, that regions above the x axis have "positive" area and regions below the x axis have "negative" area. You must account for these positive and negative values when finding the area under curves. Follow these steps.

1. Determine the x values that will serve as the left and right boundaries of the region.
2. Find all x values between the boundaries that are either solutions to the function or are values which are not in the domain of the function. These numbers are the interval numbers.
3. Integrate the function.
4. Evaluate the integral once for each of the intervals using the boundary numbers.
5. If any of the intervals evaluates to a negative number, make it positive (the negative simply tells you that the region is below the x axis).
6. Add the value of each integral to arrive at the area under the curve.

Example:

Find the area under the following function on the given intervals.
$f(x) = \sin x$; $(0, 2\pi)$

$\sin x = 0$ Find any roots to $f(x)$ on $(0, 2\pi)$.
$x = \pi$
$(0, \pi)$ $(\pi, 2\pi)$ Determine the intervals using the boundary numbers and the roots.

$\int \sin x \, dx = ^{-}\cos x$ Integrate $f(x)$. We can ignore the constant c because we have numbers to use to evaluate the integral.

$^{-}\cos x \Big]_{x=0}^{x=\pi} = {}^{-}\cos \pi - ({}^{-}\cos 0)$

$^{-}\cos x \Big]_{x=0}^{x=\pi} = {}^{-}(-1) + (1) = 2$

$^{-}\cos x \Big]_{x=\pi}^{x=2\pi} = {}^{-}\cos 2\pi - ({}^{-}\cos \pi)$

$^{-}\cos x \Big]_{x=\pi}^{x=2\pi} = {}^{-}1 + ({}^{-}1) = {}^{-}2$ The $^{-}2$ means that for $(\pi, 2\pi)$, the region is below the x axis, but the area is still 2.

Area $= 2 + 2 = 4$ Add the 2 integrals together to get the area.

Finding the **area between two curves** is much the same as finding the area under one curve. But instead of finding the roots of the functions, you need to find the x values which produce the same number from both functions (set the functions equal and solve). Use these numbers and the given boundaries to write the intervals. On each interval you must pick sample values to determine which function is "on top" of the other. Find the integral of each function. For each interval, subtract the "bottom" integral from the "top" integral. Use the interval numbers to evaluate each of these differences. Add the evaluated integrals to get the total area between the curves.

Example:

Find the area of the regions bounded by the two functions on the indicated intervals.

$f(x) = x + 2$ and $g(x) = x^2$ $[{-2}, 3]$ Set the functions equal and solve.

$x + 2 = x^2$
$0 = (x - 2)(x + 1)$
$x = 2$ or $x = {-1}$ Use the solutions and the boundary numbers to write the intervals.

$({-2}, {-1})$ $({-1}, 2)$ $(2, 3)$

$f({-3/2}) = \left(\dfrac{-3}{2}\right) + 2 = \dfrac{1}{2}$

Pick sample values on the integral and valuate each function as that number.

$g({-3/2}) = \left(\dfrac{-3}{2}\right)^2 = \dfrac{9}{4}$

$g(x)$ is "on top" on $[{-2}, {-1}]$.

$f(0) = 2$ $f(x)$ is "on top" on $[{-1}, 2]$.

$g(0) = 0$

$f(5/2) = \dfrac{5}{2} + 2 = \dfrac{9}{2}$ $g(x)$ is "on top" on $[2, 3]$.

$$g(5/2) = \left(\frac{5}{2}\right)^2 = \frac{25}{4}$$

$$\int f(x)dx = \int (x+2)dx$$

$$\int f(x)dx = \int x\,dx + 2\int dx$$

$$\int f(x)dx = \frac{1}{1+1}x^{1+1} + 2x$$

$$\int f(x)dx = \frac{1}{2}x^2 + 2x$$

$$\int g(x)dx = \int x^2 dx$$

$$\int g(x)dx = \frac{1}{2+1}x^{2+1} = \frac{1}{3}x^3$$

Area 1 = $\int g(x)dx - \int f(x)dx$ $g(x)$ is "on top" on $[^-2,^-1]$.

$$\text{Area 1} = \frac{1}{3}x^3 - \left(\frac{1}{2}x^2 + 2x\right)\Big]_{-2}^{-1}$$

$$\text{Area 1} = \left[\frac{1}{3}(^-1)^3 - \left(\frac{1}{2}(^-1)^2 + 2(^-1)\right)\right] - \left[\frac{1}{3}(^-2)^3 - \left(\frac{1}{2}(^-2)^2 + 2(^-2)\right)\right]$$

$$\text{Area 1} = \left[\frac{-1}{3} - \left(\frac{-3}{2}\right)\right] - \left[\frac{-8}{3} - (^-2)\right]$$

$$\text{Area 1} = \left(\frac{7}{6}\right) - \left(\frac{-2}{3}\right) = \frac{11}{6}$$

Area 2 = $\int f(x)dx - \int g(x)dx$ $f(x)$ is "on top" on $[^-1,2]$.

$$\text{Area 2} = \frac{1}{2}x^2 + 2x - \frac{1}{3}x^3\Big]_{-1}^{2}$$

$$\text{Area 2} = \left(\frac{1}{2}(2)^2 + 2(2) - \frac{1}{3}(2)^3\right) - \left(\frac{1}{2}(^-1)^2 + 2(^-1) - \frac{1}{3}(^-1)^3\right)$$

$$\text{Area 2} = \left(\frac{10}{3}\right) - \left(\frac{1}{2} - 2 + \frac{1}{3}\right)$$

$$\text{Area 2} = \frac{27}{6}$$

Mathematics 6-12

Area 3 = $\int g(x)dx - \int f(x)dx$ $g(x)$ is "on top" on [2,3].

Area 3 = $\frac{1}{3}x^3 - \left(\frac{1}{2}x^2 + 2x\right)\Big]_2^3$

Area 3 = $\left[\frac{1}{3}(3)^3 - \left(\frac{1}{2}(3^2) + 2(3)\right)\right] - \left[\frac{1}{3}(2)^3 - \left(\frac{1}{2}(2)^2 + 2(2)\right)\right]$

Area 3 = $\left(\frac{-3}{2}\right) - \left(\frac{-10}{3}\right) = \frac{11}{6}$

Total area = $\frac{11}{6} + \frac{27}{6} + \frac{11}{6} = \frac{49}{6} = 8\frac{1}{6}$

If you take the area bounded by a curve or curves and revolve it about a line, the result is a solid of revolution. To find the **volume** of such a solid, the Washer Method works in most instances. Imagine slicing through the solid perpendicular to the line of revolution. The "slice" should resemble a washer. Use an integral and the formula for the volume of disk.

$$\text{Volume}_{disk} = \pi \cdot radius^2 \cdot thickness$$

Depending on the situation, the radius is the distance from the line of revolution to the curve; or if there are two curves involved, the radius is the difference between the two functions. The thickness is *dx* if the line of revolution is parallel to the *x* axis and *dy* if the line of revolution is parallel to the *y* axis. Finally, integrate the volume expression using the boundary numbers from the interval.

Example:

Find the value of the solid of revolution found by revolving $f(x) = 9 - x^2$ about the x axis on the interval $[0,4]$.

radius $= 9 - x^2$
thickness $= dx$

Volume $= \int_0^4 \pi(9-x^2)^2 dx$ Use the formula for volume of a disk.

Volume $= \pi \int_0^4 (81 - 18x^2 + x^4) dx$

Volume $= \pi \left(81x - \dfrac{18}{2+1}x^3 + \dfrac{1}{4+1}x^5 \right) \Big]_0^4$ Take the integral.

Volume $= \pi \left(81x - 6x^3 + \dfrac{1}{5}x^5 \right) \Big]_0^4$

Evaluate the integral first $x = 4$ then at $x = 0$

Volume $= \pi \left[\left(324 - 384 + \dfrac{1024}{5} \right) - (0 - 0 + 0) \right]$

Volume $= \pi \left(144 \dfrac{4}{5} \right) = 144 \dfrac{4}{5} \pi = 454.9$

SKILL 9.15 Evaluate an integral by use of the fundamental theorem of calculus.

An integral is almost the same thing as an antiderivative, the only difference is the notation.

$\int_{-2}^{1} 2x\,dx$ is the integral form of the antiderivative of $2x$. The numbers at the top and bottom of the integral sign (1 and $^-2$) are the numbers used to find the exact value of this integral. If these numbers are used the integral is said to be *definite* and does not have an unknown constant c in the answer.

The fundamental theorem of calculus states that an integral such as the one above is equal to the antiderivative of the function inside (here $2x$) evaluated from $x = {}^-2$ to $x = 1$. To do this, follow these steps.

1. Take the antiderivative of the function inside the integral.
2. Plug in the upper number (here $x = 1$) and plug in the lower number (here $x = {}^-2$), giving two expressions.
3. Subtract the second expression from the first to achieve the integral value.

Examples:

1. $\int_{-2}^{1} 2x\,dx = x^2 \Big]_{-2}^{1}$ Take the antiderivative of.

   $\int_{-2}^{1} 2x\,dx = 1^2 - (^-2)^2$ Substitute in $x = 1$ and $x = {}^-2$ and subtract the results.

   $\int_{-2}^{1} 2x\,dx = 1 - 4 = {}^- 3$ The integral has the value $^-3$.

2. $\int_{0}^{\pi/2} \cos x\,dx = \sin x \Big]_{0}^{\pi/2}$

   The antiderivative of $\cos x$ is $\sin x$.

   $\int_{0}^{\pi/2} \cos x\,dx = \sin\frac{\pi}{2} - \sin 0$ Substitute in $x = \frac{\pi}{2}$ and $x = 0$. Subtract the results.

   $\int_{0}^{\pi/2} \cos x\,dx = 1 - 0 = 1$ The integral has the value 1.

COMPETENCY 10.0 KNOWLEDGE OF NUMBER SENSE AND MATHEMATICAL STRUCTURE

SKILL 10.1 Apply the properties of real numbers: closure, commutative, associative, distributive, identities, and inverses.

The real number properties are best explained in terms of a small set of numbers. For each property, a given set will be provided.

Axioms of Addition

Closure—For all real numbers a and b, $a + b$ is a unique real number.

Associative—For all real numbers a, b, and c, $(a + b) + c = a + (b + c)$.

Additive Identity—There exists a unique real number 0 (zero) such that $a + 0 = 0 + a = a$ for every real number a.

Additive Inverses—For each real number a, there exists a real number $-a$ (the opposite of a) such that $a + (-a) = (-a) + a = 0$.

Commutative—For all real numbers a and b, $a + b = b + a$.

Axioms of Multiplication

Closure—For all real numbers a and b, ab is a unique real number.

Associative—For all real numbers a, b, and c, $(ab)c = a(bc)$.

Multiplicative Identity—There exists a unique nonzero real number 1 (one) such that $1 \cdot a = a \cdot 1$.

Multiplicative Inverses—For each nonzero real number, there exists a real number $1/a$ (the reciprocal of a) such that $a(1/a) = (1/a)a = 1$.

Commutative—For all real numbers a and b, $ab = ba$.

The Distributive Axiom of Multiplication over Addition

For all real numbers a, b, and c, $a(b + c) = ab + ac$.

SKILL 10.2 Distinguish relationships between the complex number system and its subsystems.

Complex numbers are of the form $a + bi$, where a and b are real numbers and $i = \sqrt{-1}$. When i appears in an answer, it is acceptable unless it is in a denominator. When i^2 appears in a problem, it is always replaced by a $^-1$. Remember, $i^2 = {^-1}$.

To add or subtract complex numbers, add or subtract the real parts then add or subtract the imaginary parts and keep the i (just like combining like terms).

<u>Examples</u>: Add $(2 + 3i) + (^-7 - 4i)$.

$2 + {^-7} = {^-5} \qquad 3i + {^-4i} = {^-i}$ so,

$(2 + 3i) + (^-7 - 4i) = {^-5} - i$

Subtract $(8 - 5i) - (^-3 + 7i)$

$8 - 5i + 3 - 7i = 11 - 12i$

To multiply 2 complex numbers, F.O.I.L. the 2 numbers together. Replace i^2 with a $^-1$ and finish combining like terms. Answers should have the form $a + bi$.

<u>Example</u>: Multiply $(8 + 3i)(6 - 2i)$ F.O.I.L. this.

$48 - 16i + 18i - 6i^2 \qquad$ Let $i^2 = {^-1}$.

$48 - 16i + 18i - 6(^-1)$

$48 - 16i + 18i + 6$

$54 + 2i \qquad\qquad$ This is the answer.

Example: Multiply $(5+8i)^2$ ← Write this out twice.
$(5+8i)(5+8i)$ F.O.I.L. this
$25+40i+40i+64i^2$ Let $i^2 = {}^-1$.
$25+40i+40i+64({}^-1)$
$25+40i+40i-64$
${}^-39+80i$ This is the answer.

When dividing 2 complex numbers, you must eliminate the complex number in the denominator.

If the complex number in the denominator is of the form $b\,i$, multiply both the numerator and denominator by i. Remember to replace i^2 with $_-1$ and then continue simplifying the fraction.

Example:
$$\frac{2+3i}{5i}$$ Multiply this by $\frac{i}{i}$

$$\frac{2+3i}{5i} \times \frac{i}{i} = \frac{(2+3i)\,i}{5i \cdot i} = \frac{2i+3i^2}{5i^2} = \frac{2i+3({}^-1)}{{}^-5} = \frac{{}^-3+2i}{{}^-5} = \frac{3-2i}{5}$$

If the complex number in the denominator is of the form $a+b\,i$, multiply both the numerator and denominator by **the conjugate of the denominator. The conjugate of the denominator** is the same 2 terms with the opposite sign between the 2 terms (the real term does not change signs). The conjugate of $2-3i$ is $2+3i$. The conjugate of ${}^-6+11i$ is ${}^-6-11i$. Multiply together the factors on the top and bottom of the fraction. Remember to replace i^2 with $_-1$, combine like terms, and then continue simplifying the fraction.

Example:
$$\frac{4+7i}{6-5i}$$ Multiply by $\frac{6+5i}{6+5i}$, the conjugate.

$$\frac{(4+7i)}{(6-5i)} \times \frac{(6+5i)}{(6+5i)} = \frac{24+20i+42i+35i^2}{36+30i-30i-25i^2} = \frac{24+62i+35({}^-1)}{36-25({}^-1)} = \frac{{}^-11+62i}{61}$$

TEACHER CERTIFICATION STUDY GUIDE

Example:

$\dfrac{24}{-3-5i}$ Multiply by $\dfrac{-3+5i}{-3+5i}$, the conjugate.

$\dfrac{24}{-3-5i} \times \dfrac{-3+5i}{-3+5i} = \dfrac{-72+120i}{9-25i^2} = \dfrac{-72+120i}{9+25} = \dfrac{-72+120i}{34} = \dfrac{-36+60i}{17}$

Divided everything by 2.

SKILL 10.3 Apply inverse operations to solve problems (e.g., roots vs. powers, exponents vs. logarithms).

Subtraction is the inverse of Addition, and vice-versa.
Division is the inverse of Multiplication, and vice-versa.
Taking a square root is the inverse of squaring, and vice-versa.

When changing common logarithms to exponential form,

$$y = \log_b x \quad \text{if and only if} \quad x = b^y$$

Natural logarithms can be changed to exponential form by using,

$$\log_e x = \ln x \quad \text{or} \quad \ln x = y \text{ can be written as } e^y = x$$

These inverse operations are used when solving equations.

SKILL 10.4 Apply number theory concepts (e.g., primes, factors, multiples) in real-world and mathematical problem situations.

Prime numbers are numbers that can only be factored into 1 and the number itself. When factoring into prime factors, all the factors must be numbers that cannot be factored again (without using 1). Initially numbers can be factored into any 2 factors. Check each resulting factor to see if it can be factored again. Continue factoring until all remaining factors are prime. This is the list of prime factors. Regardless of what way the original number was factored, the final list of prime factors will always be the same.

Example: Factor 30 into prime factors.

    Factor 30 into any 2 factors.
        5 · 6   Now factor the 6.
        5 · 2 · 3   These are all prime factors.

    Factor 30 into any 2 factors.
        3 · 10   Now factor the 10.
        3 · 2 · 5   These are the same prime factors even though the original factors were different.

Example: Factor 240 into prime factors.

    Factor 240 into any 2 factors.
        24 · 10   Now factor both 24 and 10.
        4 · 6 · 2 · 5   Now factor both 4 and 6.
        2 · 2 · 2 · 3 · 2 · 5   These are prime factors.

This can also be written as $2^4 \cdot 3 \cdot 5$.

GCF is the abbreviation for the **greatest common factor**. The GCF is the largest number that is a factor of all the numbers given in a problem. The GCF can be no larger than the smallest number given in the problem. If no other number is a common factor, then the GCF will be the number 1. To find the GCF, list all possible factors of the smallest number given (include the number itself). Starting with the largest factor (which is the number itself), determine if it is also a factor of all the other given numbers. If so, that is the GCF. If that factor doesn't work, try the same method on the next smaller factor. Continue until a common factor is found. That is the GCF. Note: There can be other common factors besides the GCF.

Example: Find the GCF of 12, 20, and 36.

The smallest number in the problem is 12. The factors of 12 are 1,2,3,4,6 and 12. 12 is the largest factor, but it does not divide evenly into 20. Neither does 6, but 4 will divide into both 20 and 36 evenly.

Therefore, 4 is the GCF.

Example: Find the GCF of 14 and 15.

Factors of 14 are 1,2,7 and 14. 14 is the largest factor, but it does not divide evenly into 15. Neither does 7 or 2. Therefore, the only factor common to oth 14 and 15 is the number 1, the GCF.

LCM is the abbreviation for **least common multiple**. The least common multiple of a group of numbers is the smallest number that all of the given numbers will divide into. The least common multiple will always be the largest of the given numbers or a multiple of the largest number.

Example: Find the LCM of 20, 30 and 40.

The largest number given is 40, but 30 will not divide evenly into 40. The next multiple of 40 is 80 (2 x 40), but 30 will not divide evenly into 80 either. The next multiple of 40 is 120. 120 is divisible by both 20 and 30, so 120 is the LCM (least common multiple).

Example: Find the LCM of 96, 16 and 24.
The largest number is 96. 96 is divisible by both 16 and 24, so 96 is the LCM.

Divisibility Tests and Divisors

a. A number is divisible by 2 if that number is an even number (which means it ends in 0,2,4,6 or 8).
1,354 ends in 4, so it is divisible by 2. 240,685 ends in a 5, so it is not divisible by 2.

b. A number is divisible by 3 if the sum of its digits is evenly divisible by 3.

The sum of the digits of 964 is 9+6+4 = 19. Since 19 is not divisible by 3, neither is 964. The digits of 86,514 is 8+6+5+1+4 = 24. Since 24 is divisible by 3, 86,514 is also divisible by 3.

c. A number is divisible by 4 if the number in its last 2 digits is evenly divisible by 4.

The number 113,336 ends with the number 36 in the last 2 columns. Since 36 is divisible by 4, then 113,336 is also divisible by 4.

The number 135,627 ends with the number 27 in the last 2 columns. Since 27 is not evenly divisible by 4, then 135,627 is also not divisible by 4.

d. A number is divisible by 5 if the number ends in either a 5 or a 0.

225 ends with a 5 so it is divisible by 5. The number 470 is also divisible by 5 because its last digit is a 0. 2,358 is not divisible by 5 because its last digit is an 8, not a 5 or a 0.

e. A number is divisible by 6 if the number is even and the sum of its digits is evenly divisible by 3.

4,950 is an even number and its digits add to 18. (4+9+5+0 = 18) Since the number is even and the sum of its digits is 18 (which is divisible by 3), then 4950 is divisible by 6. 326 is an even number, but its digits add up to 11. Since 11 is not divisible by 3, then 326 is not divisible by 6. 698,135 is not an even number, so it cannot possibly be divided evenly by 6.

f. A number is divisible by 8 if the number in its last 3 digits is evenly divisible by 8.

The number 113,336 ends with the 3-digit number 336 in the last 3 places. Since 336 is divisible by 8, then 113,336 is also divisible by 8. The number 465,627 ends with the number 627 in the last 3 places. Since 627 is not evenly divisible by 8, then 465,627 is also not divisible by 8.

g. A number is divisible by 9 if the sum of its digits is evenly divisible by 9.

The sum of the digits of 874 is 8+7+4 = 19. Since 19 is not divisible by 9, neither is 874. The digits of 116,514 is 1+1+6+5+1+4 = 18. Since 18 is divisible by 9, 116,514 is also divisible by 9.

h. A number is divisible by 10 if the number ends in the digit 0.

305 ends with a 5 so it is not divisible by 10. The number 2,030,270 is divisible by 10 because its last digit is a 0. 42,978 is not divisible by 10 because its last digit is an 8, not a 0.

i. Why these rules work.

All even numbers are divisible by 2 by definition. A 2-digit number (with T as the tens digit and U as the ones digit) has as its sum of the digits, T + U. Suppose this sum of T + U is divisible by 3. Then it equals 3 times some constant, K. So, T + U = 3K. Solving this for U, U = 3K - T. The original 2 digit number would be represented by 10T + U. Substituting 3K - T in place of U, this 2-digit number becomes 10T + U = 10T + (3K - T) = 9T + 3K. This 2-digit number is clearly divisible by 3, since each term is divisible by 3. Therefore, if the sum of the digits of a number is divisible by 3, then the number itself is also divisible by 3. Since 4 divides evenly into 100, 200, or 300, 4 will divide evenly into any amount of hundreds. The only part of a number that determines if 4 will divide into it evenly is the number in the last 2 places. Numbers divisible by 5 end in 5 or 0. This is clear if you look at the answers to the multiplication table for 5. Answers to the multiplication table for 6 are all even numbers. Since 6 factors into 2 times 3, the divisibility rules for 2 and 3 must both work. Any number of thousands is divisible by 8. Only the last 3 places of the number determine whether or not it is divisible by 8. A 2 digit number (with T as the tens digit and U as the ones digit) has as its sum of the digits, T + U. Suppose this sum of T + U is divisible by 9. Then it equals 9 times some constant, K. So, T + U = 9K. Solving this for U, U = 9K - T. The original 2-digit number would be represented by 10T + U. Substituting 9K - T in place of U, this 2-digit number becomes 10T + U = 10T + (9K - T) = 9T + 9K.

This 2-digit number is clearly divisible by 9, since each term is divisible by 9. Therefore, if the sum of the digits of a number is divisible by 9, then the number itself is also divisible by 9. Numbers divisible by 10 must be multiples of 10 which all end in a zero.

SKILL 10.5 Identify numbers written in scientific notation, including the format used on scientific calculators and computers.

To change a number into scientific notation, move the decimal point so that only one number from 1 to 9 is in front of the decimal point. Drop off any trailing zeros. Multiply this number times 10 to a power. The power is the number of positions that the decimal point is moved. The power is negative if the original number is a decimal number between 1 and -1. Otherwise the power is positive.

Example: Change into scientific notation:

4,380,000,000	Move decimal behind the 4
4.38	Drop trailing zeros.
$4.38 \times 10^?$	Count positions that the decimal point has moved.
4.38×10^9	This is the answer.
$^-.0000407$	Move decimal behind the 4
$^-4.07$	Count positions that the decimal point has moved.
$^-4.07 \times 10^{-5}$	Note negative exponent.

If a number is already in scientific notation, it can be changed back into the regular decimal form. If the exponent on the number 10 is negative, move the decimal point to the left. If the exponent on the number 10 is positive, move the decimal point to the right that number of places.

Example: Change back into decimal form:

3.448×10^{-2} — Move decimal point 2 places left, since exponent is negative.
.03448 — This is the answer.
6×10^4 — Move decimal point 4 places right, since exponent is negative.
60,000 — This is the answer.

To add or subtract in scientific notation, the exponents must be the same. Then add the decimal portions, keeping the power of 10 the same. Then move the decimal point and adjust the exponent to keep the number in front of the decimal point from 1 to 9.

Example:

6.22×10^3
$+ 7.48 \times 10^3$ — Add these as is.
13.70×10^3 — Now move decimal 1 more place to the left and
1.37×10^4 — add 1 more exponent.

To multiply or divide in scientific notation, multiply or divide the decimal part of the numbers. In multiplication, add the exponents of 10. In division, subtract the exponents of 10. Then move the decimal point and adjust the exponent to keep the number in front of the decimal point from 1 to 9.

Example:

$(5.2 \times 10^5)(3.5 \times 10^2)$ — Multiply $5.2 \cdot 3.5$
18.2×10^7 — Add exponent
1.82×10^8 — Move decimal point and increase the exponent by 1.

Example:

$$\frac{(4.1076 \times 10^3)}{2.8 \times 10^{-4}}$$

Divide 4.1076 by 2.8

Subtract $3 - (^-4)$

1.467×10^7

COMPETENCY 11.0 KNOWLEDGE OF MATHEMATICS AS COMMUNICATION

SKILL 11.1 Identify statements that correctly communicate mathematical definitions or concepts.

In order to understand mathematics and solve problems, one must know the definitions of basic mathematic terms and concepts. For a list of definitions and explanations of basic math terms, visit the following website:

http://home.blarg.net/~math/deflist.html

Additionally, one must use the language of mathematics correctly and precisely to accurately communicate concepts and ideas.

For example, the statement "minus ten times minus five equals plus fifty" is incorrect because minus and plus are arithmetic operations not numerical modifiers. The statement should read "negative ten times negative five equals positive 50".

SKILL 11.2 Interpret written presentations of mathematics.

Because mathematics problems and concepts are often presented in written form students must have the ability to interpret written presentations and reproduce the concepts in symbolic form to facilitate manipulation and problem solving. Correct interpretation requires a sound understanding of the vocabulary of mathematics.

There are many types of written presentations of mathematics and the following are but two examples.

Examples:

1. The square of the hypotenuse of a right triangle is equivalent to the sum of the squares of the two legs.

$$a^2 + b^2 = c^2$$

2. Find the velocity of an object at time t given the objects position function is $f(t) = t^2 - 8t + 9$.

> The velocity at a given time (t) is equal to the value of the derivative of the position function at t.
> $v(t) = f'(t) = 2t - 8$
> The velocity after t seconds is $2t - 8$.

SKILL 11.3 Select or interpret appropriate concrete examples, pictorial illustrations, and symbolic representations in developing mathematical concepts.

Examples, illustrations, and symbolic representations are useful tools in explaining and understanding mathematical concepts. The ability to create examples and alternative methods of expression allows students to solve real world problems and better communicate their thoughts.

Concrete examples are real world applications of mathematical concepts. For example, measuring the shadow produced by a tree or building is a real world application of trigonometric functions, acceleration or velocity of a car is an application of derivatives, and finding the volume or area of a swimming pool is a real world application of geometric principles.

Pictorial illustrations of mathematic concepts help clarify difficult ideas and simplify problem solving.

Examples:

1. Rectangle R represents the 300 students in School A. Circle P represents the 150 students that participated in band. Circle Q represents the 170 students that participated in a sport. 70 students participated in both band and a sport.

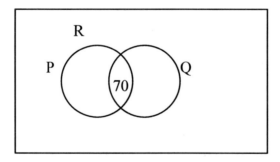

Pictorial representation of above situation.

2. A ball rolls up an incline and rolls back to its original position. Create a graph of the velocity of the ball.

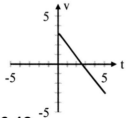

Velocity starts out at its maximum as the ball begins to roll, decreases to zero at the top of the incline, and returns to the maximum in the opposite direction at the bottom of the incline.

Symbolic representation is the basic language of mathematics. Converting data to symbols allows for easy manipulation and problem solving. Students should have the ability to recognize what the symbolic notation represents and convert information into symbolic form. For example, from the graph of a line, students should have the ability to determine the slope and intercepts and derive the line's equation from the observed data. Another possible application of symbolic representation is the formulation of algebraic expressions and relations from data presented in word problem form.

COMPETENCY 12.0 KNOWLEDGE OF MATHEMATICS AS REASONING

SKILL 12.1 Identify reasonable conjectures.

Estimation and approximation may be used to check the reasonableness of answers.

Example: Estimate the answer.

$$\frac{58 \times 810}{1989}$$

58 becomes 60, 810 becomes 800 and 1989 becomes 2000.

$$\frac{60 \times 800}{2000} = 24$$

Word problems: An estimate may sometimes be all that is needed to solve a problem.

Example: Janet goes into a store to purchase a CD on sale for $13.95. While shopping, she sees two pairs of shoes, prices $19.95 and $14.50. She only has $50. Can she purchase everything?

Solve by rounding:

$19.95→$20.00
$14.50→$15.00
$13.95→$14.00
$49.00   Yes, she can purchase the CD and the shoes.

SKILL 12.2 Identify a counter example to a conjecture.

A counterexample is an exception to a proposed rule or conjecture that disproves the conjecture. For example, the existence of a single non-brown dog disproves the conjecture "all dogs are brown". Thus, any non-brown dog is a counterexample.

In searching for mathematic counterexamples, one should consider extreme cases near the ends of the domain of an experiment and special cases where an additional property is introduced.

Examples of extreme cases are numbers near zero and obtuse triangles that are nearly flat. An example of a special case for a problem involving rectangles is a square because a square is a rectangle with the additional property of symmetry.

Example:

Identify a counterexample for the following conjectures.

1. If n is an even number, then $n+1$ is divisible by 3.

    $n = 4$
    $n + 1 = 4 + 1 = 5$
    5 is not divisible by 3.

2. If n is divisible by 3, then $n^2 - 1$ is divisible by 4.

    $n = 6$
    $n^2 - 1 = 6^2 - 1 = 35$
    35 is not divisible by 4.

SKILL 12.3 Identify simple valid arguments according to the laws of logic.

Conditional statements are frequently written in "**if-then**" form. The "if" clause of the conditional is known as the **hypothesis**, and the "then" clause is called the **conclusion**.

In a proof, the hypothesis is the information that is assumed to be true, while the conclusion is what is to be proven true. A conditional is considered to be of the form:

      If p, then q
p is the hypothesis. q is the conclusion.

Conditional statements can be diagrammed using a **Venn diagram**. A diagram can be drawn with one circle inside another circle. The inner circle represents the hypothesis. The outer circle represents the conclusion. If the hypothesis is taken to be true, then you are located inside the inner circle. If you are located in the inner circle then you are also inside the outer circle, so that proves the conclusion is true.

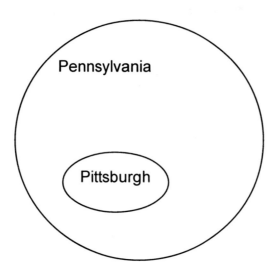

Example:
If an angle has a measure of 90 degrees, then it is a right angle.

> In this statement "an angle has a measure of 90 degrees" is the hypothesis.
> In this statement "it is a right angle" is the conclusion.

Example:
If you are in Pittsburgh, then you are in Pennsylvania.
> In this statement "you are in Pittsburgh" is the hypothesis.
> In this statement "you are in Pennsylvania" is the conclusion.

* * *

The **questioning technique** is a mathematic process skill in which students devise questions to clarify the problem, eliminate possible solutions, and simplify the problem solving process. By developing and attempting to answer simple questions, students can tackle difficult and complex problems.

Observation-inference is another mathematic process skill that is used regularly in statistics. We can use the data gathered or observed from a sample of the population to make inferences about traits and qualities of the population as a whole. For example, if we observe that 40% of voters in our sample favors Candidate A, then we can infer that 40% of the entire voting population favors Candidate A. Successful use of observation-inference depends on accurate observation and representative sampling.

SKILL 12.4 Identify proofs for mathematical assertions, including direct and indirect proofs, proofs by mathematical induction, and proofs on a coordinate plane.

Direct Proofs

In a 2 column proof, the left side of the proof should be the given information, or statements that could be proved by deductive reasoning. The right column of the proof consists of the reasons used to determine that each statement to the left was verifiably true. The right side can identify given information, or state theorems, postulates, definitions or algebraic properties used to prove that particular line of the proof is true.

Indirect Proofs

Assume the opposite of the conclusion. Keep your hypothesis and given information the same. Proceed to develop the steps of the proof, looking for a statement that contradicts your original assumption or some other known fact. This contradiction indicates that the assumption you made at the beginning of the proof was incorrect; therefore, the original conclusion has to be true.

Proofs by mathematical induction

Proof by induction states that a statement is true for all numbers if the following two statements can be proven:

1. The statement is true for $n = 1$.
2. If the statement is true for $n = k$, then it is also true for $n = k+1$.

In other words, we must show that the statement is true for a particular value and then we can assume it is true for another, larger value (k). Then, if we can show that the number after the assumed value ($k+1$) also satisfies the statement, we can assume, by induction, that the statement is true for all numbers.

The four basic components of induction proofs are: (1) the statement to be proved, (2) the beginning step ("let $n = 1$"), (3) the assumption step ("let $n = k$ and assume the statement is true for k, and (4) the induction step ("let $n = k+1$").

Example:

Prove that the sum all numbers from 1 to n is equal to $\frac{(n)(n+1)}{2}$.

Let $n = 1$.	Beginning step.
Then the sum of 1 to 1 is 1.	
And $\frac{(n)(n+1)}{2} = 1$.	
Thus, the statement is true for $n = 1$.	Statement is true in a particular instance.

Assumption

Let $n = k + 1$
$k = n - 1$

$$\text{Then } [1 + 2 + \ldots + k] + (k+1) = \frac{(k)(k+1)}{2} + (k+1) \quad \text{Substitute the assumption.}$$

$$= \frac{(k)(k+1)}{2} + \frac{2(k+1)}{2} \quad \text{Common denominator.}$$

$$= \frac{(k)(k+1) + 2(k+1)}{2} \quad \text{Add fractions.}$$

$$= \frac{(k+2)(k+1)}{2} \quad \text{Simplify.}$$

$$= \frac{(k+1)+1)(k+1)}{2} \quad \text{Write in terms of } k+1.$$

For $n = 4$, $k = 3$

$$= \frac{(4+1)(4)}{2} = \frac{20}{2} = 10$$

Conclude that the original statement is true for $n = k+1$ if it is assumed that the statement is true for $n = k$.

Proofs on a coordinate plane

Use proofs on the coordinate plane to prove properties of geometric figures. Coordinate proofs often utilize formulas such as the Distance Formula, Midpoint Formula, and the Slope Formula.

The most important step in coordinate proofs is the placement of the figure on the plane. Place the figure in such a way to make the mathematical calculations as simple as possible.

Example:

1. Prove that the square of the length of the hypotenuse of triangle ABC is equal to the sum of the squares of the lengths of the legs using coordinate geometry.

Draw and label the graph.

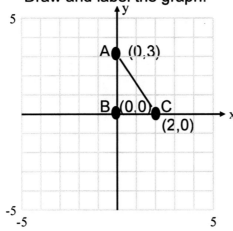

Use the distance formula to find the lengths of the sides of the triangle.

$$d = \sqrt{(x_2 - x_1)^2 + (y_2 - y_1)^2}$$

$AB = \sqrt{3^2} = 3$, $BC = \sqrt{2^2} = 2$, $AC = \sqrt{3^2 + 2^2} = \sqrt{13}$

Conclude

$(AB)^2 + (BC)^2 = 3^2 + 2^2 = 13$
$(AC)^2 = (\sqrt{13})^2 = 13$

Thus, $(AB)^2 + (BC)^2 = (AC)^2$

SKILL 12.5 Identify process skills: induction, deduction, questioning techniques, and observation-inference.

Inductive thinking is the process of finding a pattern from a group of examples. That pattern is the conclusion that this set of examples seemed to indicate. It may be a correct conclusion or it may be an incorrect conclusion because other examples may not follow the predicted pattern.

Deductive thinking is the process of arriving at a conclusion based on other statements that are all known to be true, such as theorems, axiomspostulates, or postulates. Conclusions found by deductive thinking based on true statements will **always** be true.

Examples :

Suppose:
   On Monday Mr. Peterson eats breakfast at McDonalds.
   On Tuesday Mr. Peterson eats breakfast at McDonalds.
   On Wednesday Mr. Peterson eats breakfast at McDonalds.
   On Thursday Mr. Peterson eats breakfast at McDonalds again.

Conclusion: On Friday Mr. Peterson will eat breakfast at McDonalds again.

This is a conclusion based on inductive reasoning. Based on several days observations, you conclude that Mr. Peterson will eat at McDonalds. This may or may not be true, but it is a conclusion arrived at by inductive thinking.

COMPETENCY 13.0 KNOWLEDGE OF MATHEMATICAL CONNECTIONS

SKILL 13.1 **Identify equivalent representations of the same concept or procedure (e.g., graphical, algebraic, verbal, numeric).**

Mathematical concepts and procedures can take many different forms. Students of mathematics must be able to recognize different forms of equivalent concepts.

For example, we can represent the slope of a line graphically, algebraically, verbally, and numerically. A line drawn on a coordinate plane will show the slope. In the equation of a line, $y = mx + b$, the term m represents the slope. We can define the slope of a line several different ways. The slope of a line is the change in the value of the y divided by the change in the value of x over a given interval. Alternatively, the slope of a line is the ratio of "rise" to "run" between two points. Finally, we can calculate the numeric value of the slope by using the verbal definitions and the algebraic representation of the line.

SKILL 13.2 **Interpret relationships between mathematical topics (e.g., multiplication as repeated addition, powers as repeated multiplication).**

Recognition and understanding of the relationships between concepts and topics is of great value in mathematical problem solving and the explanation of more complex processes.

For instance, multiplication is simply repeated addition. This relationship explains the concept of variable addition. We can show that the expression $4x + 3x = 7x$ is true by rewriting 4 times x and 3 times x as repeated addition, yielding the expression $(x + x + x + x) + (x + x + x)$. Thus, because of the relationship between multiplication and addition, variable addition is accomplished by coefficient addition.

Another example of a mathematical relationship is powers as repeated multiplication. This relationship explains the rules of exponent operations. For instance, the multiplication of exponential terms with like bases is accomplished by the addition of the exponents.

$$2^2 \times 2^5 = 2^7$$

$$(2 \times 2) + (2 \times 2 \times 2 \times 2 \times 2) = 2^{2+5} = 2^7$$

This rule yields the general formula for the product of exponential terms, $z^m \times z^n = z^{m+n}$, that is useful in problem solving.

SKILL 13.3 Interpret descriptions, diagrams, and representations of arithmetic operations.

Students of mathematics must be able to recognize and interpret the different representations of arithmetic operations.

First, there are many different verbal descriptions for the operations of addition, subtraction, multiplication, and division. The table below identifies several words and/or phrases that are often used to denote the different arithmetic operations.

Operation	Descriptive Words
Addition	"plus", "combine", "sum", "total", "put together"
Subtraction	"minus", "less", "take away", "difference"
Multiplication	"product", "times", "groups of"
Division	"quotient", "into", "split into equal groups",

Second, diagrams of arithmetic operations can present mathematical data in visual form. For example, we can use the number line to add and subtract.

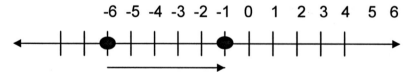

The addition of 5 to -4 on the number line; -4 + 5 = 1.

Finally, as shown in the examples below, we can use pictorial representations to explain all of the arithmetic processes.

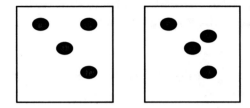

Two groups of four equals eight or 2 x 4 = 8 shown in picture form.

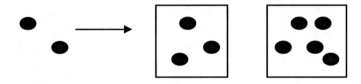

Adding three objects to two or 3 + 2 = 5 shown in picture form.

TEACHER CERTIFICATION STUDY GUIDE

COMPETENCY 14.0 KNOWLEDGE OF INSTRUCTION

SKILL 14.1 Select appropriate resources for a classroom activity (e.g., manipulatives, mathematics models, technology, other teaching tools).

The use of supplementary materials in the classroom can greatly enhance the learning experience by stimulating student interest and satisfying different learning styles. Manipulatives, models, and technology are examples of tools available to teachers.

Manipulatives are materials that students can physically handle and move. Manipulatives allow students to understand mathematic concepts by allowing them to see concrete examples of abstract processes. Manipulatives are attractive to students because they appeal to the students' visual and tactile senses. Available for all levels of math, manipulatives are useful tools for reinforcing operations and concepts. They are not, however, a substitute for the development of sound computational skills.

Models are another means of representing mathematical concepts by relating the concepts to real-world situations. Teachers must choose wisely when devising and selecting models because, to be effective, models must be applied properly. For example, a building with floors above and below ground is a good model for introducing the concept of negative numbers. It would be difficult, however, to use the building model in teaching subtraction of negative numbers.

Finally, there are many forms of **technology** available to math teachers. For example, students can test their understanding of math concepts by working on skill specific computer programs and websites. Graphing calculators can help students visualize the graphs of functions. Teachers can also enhance their lectures and classroom presentations by creating multimedia presentations.

SKILL 14.2 Identify methods and strategies for teaching problem-solving skills and applications (e.g., constructing tables from given data, guess-and-check, working backwards, reasonableness, estimation).

Successful math teachers introduce their students to multiple problem solving strategies and create a classroom environment where free thought and experimentation are encouraged.

Teachers can promote problem solving by allowing multiple attempts at problems, giving credit for reworking test or homework problems, and encouraging the sharing of ideas through class discussion. There are several specific problem solving skills with which teachers should be familiar.

The **guess-and-check** strategy calls for students to make an initial guess at the solution, check the answer, and use the outcome as a guide for the next guess. With each successive guess, the student should get closer to the correct answer. Constructing a table from the guesses can help organize the data.

Example:

There are 100 coins in a jar. 10 are dimes. The rest are pennies and nickels. There are twice as many pennies as nickels. How many pennies and nickels are in the jar?

There are 90 total nickels and pennies in the jar (100 coins – 10 dimes).

There are twice as many pennies as nickels. Make guesses that fulfill the criteria and adjust based on the answer found. Continue until we find the correct answer, 60 pennies and 30 nickels.

Number of Pennies	Number of Nickels	Total Number of Pennies and Nickels
40	20	60
80	40	120
70	35	105
60	30	90

When solving a problem where the final result and the steps to reach the result are given, students must **work backwards** to determine what the starting point must have been.

Example:

John subtracted seven from his age, and divided the result by 3. The final result was 4. What is John's age?

Work backward by reversing the operations.
4 x 3 = 12;
12 + 7 = 19
John is 19 years old.

Estimation and testing for **reasonableness** are related skills students should employ prior to and after solving a problem. These skills are particularly important when students use calculators to find answers.

Example:

Find the sum of 4387 + 7226 + 5893.

4300 + 7200 + 5800 = 17300   Estimation.
4387 + 7226 + 5893 = 17506   Actual sum.

By comparing the estimate to the actual sum, students can determine that their answer is reasonable.

TEACHER CERTIFICATION STUDY GUIDE

COMPETENCY 15.0 KNOWLEDGE OF ASSESSMENT

SKILL 15.1 Identify students' errors, including multiple errors that result in correct or incorrect answers (e.g., algorithms, properties, drawings, procedures).

Successful teachers are able to identify all types of student errors. Identifying patterns of errors allows teachers to tailor their curriculum to enhance student learning and understanding. The following is a list of common errors committed by algebra and calculus students.

<u>Algebra</u>
1. Failing to distribute:
    ex. 4x − (2x − 3) = 4x − 2x + 3 (**not** 4x − 2x − 3)
    ex. 3(2x + 5) = 6x + 15 (**not** 6x + 5)
2. Distributing exponents:
    ex. $(x + y)^2 = x^2 + 2xy + y^2$ (**not** $x^2 + y^2$)
3. Canceling terms instead of factors:

    ex. $x^2 + \frac{5}{x} \left(\text{not } x^2 + 5x^2 + 5\right) X^2 + 5 = x^2 + \frac{5}{x}\left(\text{not } x^2 + 5x^2 + 5\right)$

4. Misunderstanding negative and fractional exponents:

    ex. $x^{\frac{1}{2}} = \sqrt{x}$ (**not** $\frac{1}{x^2}$)

    ex. $x^{-2} = \frac{1}{x^2}$ (**not** $\sqrt{x}$)

<u>Calculus</u>
1. Failing to use the product, quotient, and chain rules:
    $f(x) = (5x^2)(3x^3),$
    ex. $f'(x) = (10x)(3x^3) + (5x^2)(9x^2)$
    $= 30x^4 + 45x^4 = 75x^4$
    [**not** $(10x)(9x^2) = 90x^3$]
2. Misapplying derivative and integration rules:
    ex.
    $\int \frac{1}{x} = \ln x + C,$
    $\int \frac{1}{(x^2+1)} = \frac{1}{\tan x} + C$ (**not** $\ln(x^2+1) + C$)

Teachers also must not neglect multiple errors in the work of a student that may be disguised by a correct answer. For example, failing twice to properly distribute negative one in the following arithmetic operation produces a deceptively correct answer.

$4x - (2x+3) - (2x - 3) = 4x - 2x + 3 - 2x - 3 = -2x$ (incorrect distribution)

$4x - (2x+3) - (2x - 3) = 4x - 2x - 3 - 2x + 3 = -2x$ (correct distribution)

SKILL 15.2 Identify appropriate alternative methods of assessment (e.g., performance, portfolios, projects).

In addition to the traditional methods of performance assessment like multiple choice, true/false, and matching tests, there are many other methods of student assessment available to teachers. Alternative assessment is any type of assessment in which students create a response rather than choose an answer.

Short response and **essay** questions are alternative methods of performance assessment. In responding to such questions, students must utilize verbal, graphical, and mathematical skills to construct answers to problems. These multi-faceted responses allow the teacher to examine more closely a student's problem solving and reasoning skills.

Student **portfolios** are another method of alternative assessment. In creating a portfolio, students collect samples and drafts of their work, self-assessments, and teacher evaluations over a period of time. Such a collection allows students, parents, and teachers to evaluate student progress and achievements. In addition, portfolios provide insight into a student's thought process and learning style.

Projects, **demonstrations**, and **oral presentations** are means of alternative assessment that require students to use different skills than those used on traditional tests. Such assessments require higher order thinking, creativity, and the integration of reasoning and communication skills. The use of predetermined rubrics, with specific criteria for performance assessment, is the accepted method of evaluation for projects, demonstrations, and presentations.

TEACHING METHODS - The art and science specific for high school mathematics

Some commonly used teaching techniques and tools are described below along with links to further information. The links provided provide a wealth of instructional ideas and materials. You should consider joining The National Council of Teachers of Mathematics as they have many ideas in their journals about pedagogy and curriculum standards and publish professional books that are useful. You can write to them at 1906 Association Drive, Reston, VA 20191-1593. You can also order a starter kit from them for $9 that includes 3 recent journals by calling 800-235-7566 or writing e-mailorders@nctm.org

A couple of resources for students to use at home:
http://www.algebra.com/, http://www.mathsisfun.com/algebra/index.html and http://www.purplemath.com/. A helpful .pdf guide for parents is available at:
http://my.nctm.org/ebusiness/ProductCatalog/product.aspx?ID=12931
A good website for understanding the causes of and how to prevent "math anxiety:"
http://www.mathgoodies.com/articles/math_anxiety.html

1. Classroom warm-up: Engage your students as soon as they walk in the door: provide an interesting short activity each day. You can make use of thought-provoking questions and puzzles. Also use relevant puzzles specific to topics you may be covering in your class. The following websites provide some ideas:

    http://www.math-drills.com/?gclid=CP-P0dzenJICFRSTGgodNjG0Zw
    http://www.mathgoodies.com/games/
    http://mathforum.org/k12/k12puzzles/
    http://mathforum.org/pow/other.html

2. Real life examples: Connect math to other aspects of your students' lives by using examples and data from the real world whenever possible. It will not only keep them engaged, it will also help answer the perennial question "Why do we have to learn math?" Online resources to get you started:

    http://chance.dartmouth.edu/chancewiki/index.php/Main_Page has some interesting real-world probability problems (such as, "Can statistics determine if Robert Clemens used steroids?")
    http://www.mathnotes.com/nos_index.html has all kinds of links between math and the real world suitable for high school students
    http://www.nssl.noaa.gov/edu/ideas/ uses weather to teach math
    http://standards.nctm.org/document/eexamples/index.htm#9-12 Using Graphs, Equations, and Tables to Investigate the Elimination of Medicine from the Body: Modeling the Situation

http://mathforum.org/t2t/faq/election.html Election math in the classroom
http://www.education-world.com/a_curr/curr148.shtml offers examples of real-life problems such as calculating car payments, saving and investing, the world of credit cards, and other finance problems.
http://score.kings.k12.ca.us/real.world.html is a website connecting math to real jobs, elections, NASA projects, etc.

3. Graphing and spreadsheets for enhancing math learning:
   http://www.cvgs.k12.va.us/digstats/
   http://score.kings.k12.ca.us/standards/probability.html for graphing and statistics.
   http://www.microsoft.com/education/solving.mspx for using spreadsheets and to solve polynomial problems.

4. Use technology - manipulatives, software and interactive online activities that can help all students learn, particularly those oriented more towards visual and kinesthetic learning. Here are some websites:
   http://illuminations.nctm.org/ActivitySearch.aspx has games for grades 9-12 that can be played against the computer or another student.
   http://nlvm.usu.edu/ The National Library of Virtual Manipulatives has resources for all grades on numbers and operations, algebra, geometry, probability and measurement.
   http://mathforum.org/pow/other.html has links to various math challenges, manipulatives and puzzles.
   http://www.etacuisenaire.com/algeblocks/algeblocks.jsp Algeblocks are blocks that utilize the relationship between algebra and geometry.

5. Word problem strategies- the hardest thing to do is take the English and turn it into math but there are 6 key steps to teach students how to solve word problems:

   a. The problem will have **key words** to suggest the type of operation or operations to be performed to solve the problem. For example, words such as "altogether" or "total" imply addition while words such as "difference" or "How many more?" imply subtraction.
   b. **Pictures or Concrete Materials**: Math is very abstract; it is easier to solve a problem using pictures or concrete materials to illustrate the problem. Pictures and concrete materials allow the students to manipulate the material to solve the problem with trial and error. Model drawing pictures and using concrete materials to solve word problems.

c. *Use Logic*: Ask your students if their answers make sense. Get them used to using the process of deduction. Model the deduction process for them to decide on the answer to a word problem. For example, in solving a problem such as: Two consecutive numbers have a sum of 91. What are the numbers? If the student arrives at an answer of 44 and 45 it is obvious that there was an error in the equation used or calculation since 44 and 45 are consecutive but don't add up to 91. Let x = the 1st number and (x+1) = the 2nd number, so that x + (x+1) =91 and 2x +1 =91, then 2x=90, x=45 and x+1=46. The answer is 45 and 46.
d. ***Eliminate the possibilities and look for patterns or work the problem backwards***
e. ***Guess the Answer***: Students should guess an approximate answer that makes sense based on the problem. For example, if the student knows the word problem implies addition, they should recognize that the answer must be greater than the numbers in the problem. Often students are afraid of guessing because they don't want to get the wrong answer but encourage your students to guess and then double check the answer to see if it works. If the answer is incorrect, the student can try another strategy for finding the answer.
f. ***Make a Table***: Selecting relevant information from a word problem and organizing the data is very helpful in solving word problems. Often students become confused because there are too many numbers and/or variables.

http://www.purplemath.com/modules/translat.htm, http://math.about.com/library/weekly/aa071002a.htm and http://www.onlinemathlearning.com/algebra-word-problems.html are great resources for students to solve word problems.

6. Mental math practice
   Give students regular practice in doing mental math. The following websites offer many mental calculation tips and strategies:
   http://www.cramweb.com/math/index.htm
   http://mathforum.org/k12/mathtips/mathtips.html

Because frequent calculator use tends to deprive students of a sense of numbers and an ability to calculate on their own, they will often approach a sequence of multiplications and divisions the hard way. For instance, asked to calculate 770 x 36/ 55, they will first multiply 770 and 36 and then do a long division with the 55. They fail to recognize that both 770 and 55 can be divided by 11 and then by 5 to considerably simplify the problem. Give students plenty of practice in multiplying and dividing a sequence of integers and fractions so they are comfortable with canceling top and bottom terms.

7. Math language
   Math vocabulary help is available for high school students on the web:

   http://www.amathsdictionaryforkids.com/ is a colorful website math dictionary
   http://www.math.com/tables/index.html is a math dictionary in English and Spanish

ERROR ANALYSIS

A simple method for analyzing student errors is to ask how the answer was obtained. The teacher can then determine if a common error pattern has resulted in the wrong answer. There is a value to having the students explain how the arrived at the correct as well as the incorrect answers.

Many errors are due to simple **carelessness**. Students need to be encouraged to work slowly and carefully. They should check their calculations by redoing the problem on another paper, not merely looking at the work. Addition and subtraction problems need to be written neatly so the numbers line up. Students need to be careful regrouping in subtraction. Students must write clearly and legibly, including erasing fully. Use estimation to ensure that answers make sense.

Many students' computational skills exceed their **reading** level. Although they can understand basic operations, they fail to grasp the concept or completely understand the question. Students must read directions slowly.

Fractions are often a source of many errors. Students need to be reminded to use common denominators when adding and subtracting and to always express answers in simplest terms. Again, it is helpful to check by estimating.

The most common error that is made when working with **decimals** is failure to line up the decimal points when adding or subtracting or not moving the decimal point when multiplying or dividing. Students also need to be reminded to add zeroes when necessary. Reading aloud may also be beneficial. Estimation, as always, is especially important.

Students need to know that it is okay to make mistakes. The teacher must keep a positive attitude, so they do not feel defeated or frustrated.

THE ART OF TEACHING - PEDAGOGICAL PRINCIPLES
Maintain a supportive, non-threatening environment

The key to success in teaching goes beyond your mathematical knowledge and the desire to teach. Though important, knowledge and desire alone do not make you a good teacher. Being able to connect with your students is vital: learn their names immediately, have a seating chart the first day (even if you intend to change it) and learn about your students -their hobbies, phone number, parent's names, what they like and dislike about school and learning and math. Keep this information on each student *and learn it:* adapt your lessons, how challenging they are and what other resources you may need to accommodate your students' individual strengths and weaknesses.

Learn to see math as your students see it. If you aren't able to connect with your students, no matter how well your lessons are and how well you know the material, you won't inspire them to learn math from you. As you expect respect, you must give respect and as you expect their attention, they also need your attention and understanding. Talk to them with the same tone of voice as you would an adult, not in a tone that makes them feel like children. Look your students in the eye when you talk to them and encourage questions and comments. Take advantage of teachable moments and explain the rationale behind math rules.

Demonstrate respect, care and trust toward every student; assume the best. This does not mean becoming "friends" with your students or you will have problems with discipline. You can be kind and firm at the same time. Have a fair and clear grading and discipline system that is posted, reviewed and made clear to your students. Consistency, structure and fairness are essential to earning their trust in you as a teacher. Always admit your mistakes and be available to your students certain days after school. Finally, demonstrate your care for them and your love of math and you will be a positive influence on their learning.

Below are websites to help make your teaching more effective and fun:

1. **Teachers Helping Teachers** has several resources for high school mathematics.
2. **Math Resources for Teachers –** resources for grades 7 - 10
3. **Math is Marvelous Web Site** - is a fascinating website on the history of geometry
4. **Math Archives K-12** resources for lesson plans and software
5. http://www.edhelper.com/algebra.htm covers Algebra I & II
6. **Math Goodies** interactive lessons, worksheets and homework help
7. **Multicultural Lessons** an interesting site with lessons on Babylonian Square Roots, Chinese Fraction Reducing, Egyptian multiplication, etc.
8. http://www.goenc.com/ resources and professional development for teachers
9. **Math and Reading Help** a guide to math, reading, homework help, tutoring and earning a high school diploma
10. **Purple Math.com** all about Algebra, lessons, help for students and lots of other resources

18. http://library.thinkquest.org/20991/home.html this site has Algebra, Geometry and Pre-calc/Calculus
19. http://www.math.com/ this site has Algebra, Geometry, Trigonometry, and Calculus, plus homework help
20. http://www.math.armstrong.edu/MathTutorial/index.html a tutorial in algebra
21. http://www.wtamu.edu/academic/anns/mps/math/mathlab/beg_algebra/index.htm this site is helpful for those beginning Algebra or for a refresher
22. **Math Complete** this radicals, quadratics, linear equation solvers
23. **Math Tutor - PEMDAS & Integers** fractions, intergers, information for parents and teachers
24. **Matrix Algebra** all about matrix operations and applications

25. **Mr. Stroh's Algebra Site** help for Algebra I & II
26. **Polynomials and Polynomial Functions** everything from factoring, to graphing, finding rational zeros and multiplying, adding and subtracting polynomials
27. **Quadratic Formula** all about quadratics
28. **Animated Pythagorean Theorem** a fun an animated proof of the Pythagorean Theorem
29. **Brunnermath - Interactive Activities** general math, Algebra, Geometry, Trigonometry, Statistics, Calculus, using Calculators
30. **CoolMath4Kids Geometry** creating art with math and geometry lessons
31. **The Curlicue Fractal** The curlicue fractal is an exceedingly easy-to-make but richly complex pattern using trigonometry and calculus to create fascinating shapes
32. **Euclid's Elements Interactive** Euclid's *Elements* form one of the most beautiful and influential works of science in the history of humankind.
33. **Howe-Two Free Software** software solutions for mathematics instruction
34. http://regentsprep.org/regents/math/math-topic.cfm?TopicCode=syslin Systems of equations lessons and practice
35. http://www.sparknotes.com/math/algebra1/systemsofequations/problems3.rhtml Word problems system of equations
36. http://math.about.com/od/complexnumbers/Complex_Numbers.htm Several complex number exercise pages
37. http://regentsprep.org/Regents/math/ALGEBRA/AE3/PracWord.htm practice with Systems of inequalities word problems
38. http://regentsprep.org/regents/Math/solvin/PSolvIn.htm solving inequalities
39. http://www.wtamu.edu/academic/anns/mps/math/mathlab/beg_algebra/beg_alg_tut18_ineq.htm Inequality tutorial, examples, problems
40. http://www.wtamu.edu/academic/anns/mps/math/mathlab/beg_algebra/beg_alg_tut24_ineq.htm Graphing linear inequalities tutorial

TEACHER CERTIFICATION STUDY GUIDE

41. http://www.wtamu.edu/academic/anns/mps/math/mathlab/col_algebra/col_alg_tut17_quad.htm Quadratic equations tutorial, examples, problems
42. http://regentsprep.org/Regents/math/math-topic.cfm?TopicCode=factor Practice factoring
43. http://www.wtamu.edu/academic/anns/mps/math/mathlab/col_algebra/col_alg_tut37_syndiv.htm Synthetic division tutorial
44. http://www.tpub.com/math1/10h.htm Synthetic division Examples and problems

DEVELOPMENTAL PSYCHOLOGY AND TEACHING MATHEMATICS- things you may not know about your students:

Studies show that health matters more than gender or social status when it comes to learning. Healthy girls and boys do equally well on most cognitive tasks. Boys perform better at manipulating shapes and analyzing and girls perform better on processing speed and motor dexterity. No differences have been measured in calculation ability, meaning girls and boys have an equal aptitude for mathematics.

The following was written by Jay Giedd, M.D. is a practicing Child and Adolescent Psychiatrist and Chief of Brain Imaging at the Child Psychiatry Branch of the National Institute of Mental Health:

http://nihrecord.od.nih.gov/newsletters/2005/08_12_2005/story04.htm

"The most surprising thing has been how much the teen brain is changing. By age six, the brain is already 95 percent of its adult size. But the gray matter, or thinking part of the brain, continues to thicken throughout childhood as the brain cells get extra connections, much like a tree growing extra branches, twigs and roots...In the frontal part of the brain, the part of the brain involved in judgment, organization, planning, strategizing -- those very skills that teens get better and better at -- this process of thickening of the gray matter peaks at about age 11 in girls and age 12 in boys, roughly about the same time as puberty. After that peak, the gray matter thins as the excess connections are eliminated or pruned...

Contrary to what most parents have thought at least once, "teens really do have brains," quipped Dr. Jay Giedd, NIMH intramural scientist, in a lecture on the "Teen Brain Under Construction." His talk was the kick-off event for the recent NIH Parenting Festival. Giedd said scientists have only recently learned more about the trajectories of brain growth. One of the findings he discussed showed the frontal cortex area — which governs judgment, decision-making and impulse control — doesn't fully mature until around age 25. "That really threw us," he said. "We used to joke about having to be 25 to rent a car, but there's tons of data from insurance reports [showing] that 24-year-olds are costing them more than 44-year-olds."

So why is that? "It must be behavior and impulse control," he said. "Whatever these changes are, the top 10 bad things that happen to teens involve emotion and behavior." Physically, Giedd said, the teen years and early 20s represent an incredibly healthy time of life, in terms of cancer, heart disease and other serious illnesses. But with accidents as the leading cause of death in adolescents, and suicide following close behind, "this isn't a great time emotionally and psychologically. This is the great paradox of adolescence: right at the time you should be on the top of your game, you're not."

The next step in Giedd's research, he said, is to learn more about what influences brain growth, for good or bad. "Ultimately, we want to use these findings to treat illness and enhance development."

One of the things scientists have come to understand, though, is that parents do have something to do with their children's brain development.

"From imaging studies, one of the things that seems intriguing is this notion of modeling...that the brain is pretty adept at learning by example," he said. "As parents, we teach a lot when we don't even know we're teaching, just by showing how we treat our spouses, how we treat other people, what we talk about in the car on the way home...things that a parent says in the car can stick with them for years. They're listening even though it may appear they're not."

What can we do to change our kids? "Well, start with yourself in terms of what you show by example," Giedd concluded.

Maybe the parts of the brain performing geometry are different from the parts doing algebra. There is no definitive research to answer that question yet, but it is obviously what researchers are looking for.

Time-Lapse Imaging Tracks Brain Maturation from ages 5 to 20
Constructed from MRI scans of healthy children and teens, the time-lapse "movie", from which the above images were extracted, compresses 15 years of brain development (ages 5–20) into just a few seconds.

Red indicates **more** gray matter, **blue less** gray matter. Gray matter wanes in a back-to-front wave as the brain matures and neural connections are pruned.
Source: Paul Thompson, Ph.D. UCLA Laboratory of Neuroimaging
http://www.loni.ucla.edu/%7Ethompson/DEVEL/PR.html

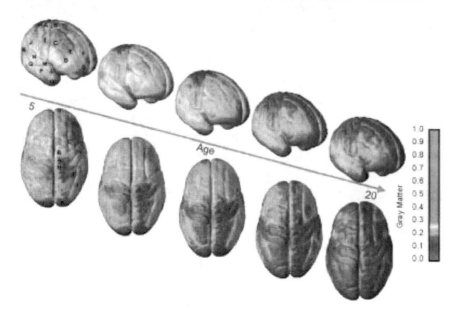

What are the implications of this fascinating study for teachers? It's unreasonable to expect teens to have adult levels of organizational skills or decision-making before their brains have completely developed. In teens, the frontal lobe, or the executive of the brain is what handles organizing, decision making, emotions, attending, shifting attention, planning and making strategies and it is not fully developed until the early to mid-twenties.

*Perhaps since certain parts of the brain develop sooner than others, subjects should be taught in a different order. Until we know more, just understanding that the parts of teen brains related to decision making and emotions are still developing through the early 20's is important, and **that stressful situations lead to diminished ability to made good judgments**. For some children, just being called on in class is stressful. At this age, social relationships become very important and **teachers need to be sensitive to this aspect of teen development as it relates to stress and decision-making**. The immaturity of this part of the teen brain might explain why the teen crash rate is 4 times that of adults...*

ANSWER KEY TO PRACTICE PROBLEMS

Skill 1.1, page 1

Question #1

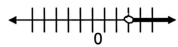

Question #2

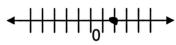

Question #3

Question #4

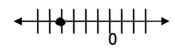

Skill 1.2, page 2

Question #1

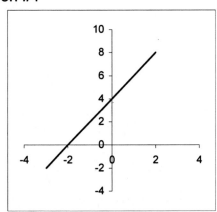

Question #2

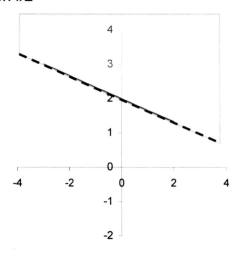

Question #3

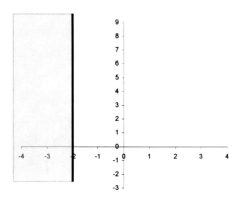

Skill 1.3, page 4

Question #1   x-intercept = -14   y-intercept = -10   slope = $-\dfrac{5}{7}$

Question #2   x-intercept = 14   y-intercept = -7   slope = $\dfrac{1}{2}$

Question #3   x-intercept = 3   y-intercept = none

Question #4   x-intercept = $\dfrac{15}{2}$ y-intercept = 3   slope = $-\dfrac{2}{5}$

Skill 1.5, page 6

Question #1 $y = \dfrac{3}{4}x + \dfrac{17}{4}$

Question #2 $x = 11$

Question #3 $y = \dfrac{3}{5}x + \dfrac{42}{5}$

Question #4 $y = 5$

page 8

Question #1 $\dfrac{8x+36}{(x-3)(x+7)}$

Question #2 $\dfrac{25a^2+12b^2}{20a^4b^5}$

Question #3 $\dfrac{2x^2+5x-21}{(x-5)(x+5)(x+3)}$

page 10

Question #1 It takes Curly 15 minutes to paint the elephant alone
Question #2 The original number is 5/15
Question #3 The car was traveling at 68mph and the truck was traveling at 62mph

page 11

Question #1 $C = \dfrac{5}{9}F - \dfrac{160}{9}$

Question #2 $b = \dfrac{2A - 2h^2}{h}$

Question #3 $n = \dfrac{360 + S}{180}$

page 12

Question #1 $x = 7,\ x = {}^-5$

Question #2 $x = \dfrac{13}{8}$

Skill 1.6, *page 14*

Question #1 $(6x - 5y)(36x^2 + 30xy + 25y^2)$
Question #2 $4(a - 2b)(a^2 + 2ab + 4b^2)$
Question #3 $5x^2(2x^9 + 3y)(4x^{18} - 6x^9y + 9y^2)$

Skill 1.7, *page 15*

Question #1 $9xz^5$

Question #2 $\dfrac{3x + 2y}{x^2 + 5xy + 25y^2}$

Question #3 $\dfrac{x^2 + 8x + 15}{(x+2)(x+3)(x-7)}$

Mathematics 6-12

page 16

    Question #1 $6a^4\sqrt{2a}$

    Question #2 $7i\sqrt{2}$

    Question #3 $-2x^2$

    Question #4 $6x^4y^3\sqrt[4]{3x^2y^3}$

Skill 1.8, *page 19*

    Question #1 $\dfrac{14x+28}{(x+6)(x+1)(x-1)}$

    Question #2 $\dfrac{x^3-5x^2+10x-12}{x^2+3x-10}$

page 20

    Question #1 $17\sqrt{6}$

    Question #2 $84x^4y^8\sqrt{2}$

    Question #3 $\dfrac{5a^4\sqrt{35b}}{8b}$

    Question #4 $-5\left(3+\sqrt{3}+2\sqrt{5}\right)$

Skill 1.9, *page 22*

    Question #1 $x=11$

    Question #2 $x=17$

Skill 1.10, *page 23*

    Question #1 $5\sqrt{6}$

    Question #2 $^-5\sqrt{3}+6\sqrt{5}=\sqrt{15}-30$

    Question #3 $^-3\sqrt{2}-9\sqrt{3}-3\sqrt{6}-12$

Skill 1.12, *page 27*

    Question #1   The sides are 8, 15, and 17

    Question #2   The numbers are 2 and $\dfrac{1}{2}$

Skill 1.13, *page 30*

    Question #1 $x=\dfrac{7\pm\sqrt{241}}{12}$

    Question #2 $x=\dfrac{1}{2}$ or $\dfrac{^-2}{5}$

    Question #3 $x=2$ or $\dfrac{6}{5}$

Skill 1.14, *page 31*

    Question #1 $x^2 - 10x + 25$
    Question #2 $25x^2 - 10x - 48$
    Question #3 $x^2 - 9x - 36$

Skill 1.22, *page 52*

    Question #1 $35x^3 y^4$
    Question #2 $78732 x^7 y^2$

Skill 2.2, *page 60*

    Question #2 a, b, c, f are functions
    Question #3 Domain = $^-\infty, \infty$ Range = $^-5, \infty$

Skill 2.3, *page 62*

    Question #1 Domain = $^-\infty, \infty$ Range = $^-6, \infty$
    Question #2 Domain = 1,4,7,6   Range = -2
    Question #3 Domain = $x \neq 2, ^-2$
    Question #4 Domain = $^-\infty, \infty$ Range = -4, 4
                     Domain = $^-\infty, \infty$ Range = $2, \infty$
    Question #5 Domain = $^-\infty, \infty$ Range = 5
    Question #6 (3,9), (-4,16), (6,3), (1,9), (1,3)

Skill 2.4, page 68

Question #1

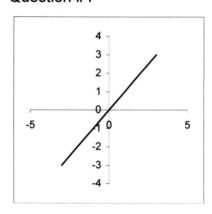

Question #2

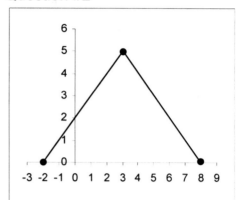

Question #3

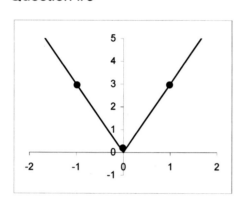

Questions #4

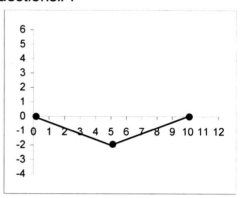

page 69

Question #1

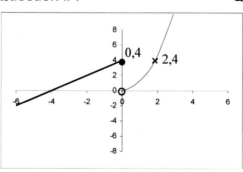

Question #2

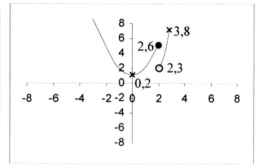

Skill 3.17, page 124

    Question #1   The Red Sox won the World Series.
    Question #2   Angle B is not between 0 and 90 degrees.
    Question #3   Annie will do well in college.
    Question #4   You are witty and charming.

Skill 5.2, *page 150*

Question #1
$$\cot\theta = \frac{x}{y}$$
$$\frac{x}{y} = \frac{x}{r} \times \frac{r}{y} = \frac{x}{y} = \cot\theta$$

Question #2
$$1 + \cot^2\theta = \csc^2\theta$$
$$\frac{y^2}{y^2} + \frac{x^2}{y^2} = \frac{r^2}{y^2} = \csc^2\theta$$

Skill 8.3, *page 185*

Question #1 $S_5 = 75$

Question #2 $S_n = 28$

Question #3 $S_n = -\frac{-31122}{15625} \approx ^-1.99$

Skill 8.5, *page 188*

Question #1 $\begin{pmatrix} 11 & 7 \\ 11 & 1 \end{pmatrix}$

Question #2 $\begin{pmatrix} 0 & 3 \\ -10 & 13 \\ 5 & 0 \end{pmatrix}$

page 189

Question #1 $\begin{pmatrix} -4 & 0 & -2 \\ 2 & 4 & -8 \end{pmatrix}$

Question #2 $\begin{pmatrix} 18 \\ 34 \\ 32 \end{pmatrix}$

Question #3 $\begin{pmatrix} -12 & 16 \\ -4 & -2 \\ 0 & 6 \end{pmatrix}$

Skill 8.6, *page 191*

Question #1 $\begin{pmatrix} -15 & 25 \\ -1 & -13 \end{pmatrix}$

Question #2 $\begin{pmatrix} 5 & -5 & -10 \\ 5 & 5 & 0 \\ 1 & 8 & 7 \\ -9 & 13 & 22 \end{pmatrix}$

page 192

Question #1 $\begin{pmatrix} x \\ y \end{pmatrix} = \begin{pmatrix} 3 \\ 1 \end{pmatrix}$

Question #2 $\begin{pmatrix} x \\ y \\ z \end{pmatrix} = \begin{pmatrix} 4 \\ 4 \\ 1 \end{pmatrix}$

Skill 9.2, *page 196*

Question # 1   49.34

Question # 2   1

Skill 9.11, *page 215*

Question #1 $t(0) = -24$ m/sec
Question #2 $t(4) = 24$ m/sec

Skill 9.13, *page 219*

Question #1 $g(x) = -20\sin x + c$
Question #2 $g(x) = \pi \sec x + c$

SAMPLE TEST

Directions: Read each item and select the best response.

1. Which graph represents the solution set for $x^2 - 5x > -6$?
   (Average Rigor) (Skill 1.1)

   A) ←―○―――○―→
          -2 0 2

   B) ←○――――○―→
         -3 0 3

   C) ←―○―――○―→
          -2 0 2

   D) ←―――――○○―→
         -3 0 2 3

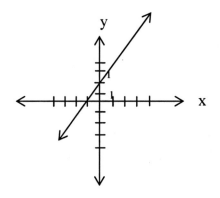

2. What is the equation of the above graph?
   (Easy) (Skill 1.2)

   A) $2x + y = 2$
   B) $2x - y = -2$
   C) $2x - y = 2$
   D) $2x + y = -2$

3. Solve for v_0: $d = at(v_t - v_0)$
   (Average Rigor) (Skill 1.5)

   A) $v_0 = atd - v_t$
   B) $v_0 = d - atv_t$
   C) $v_0 = atv_t - d$
   D) $v_0 = (atv_t - d)/at$

4. Which of the following is a factor of $6 + 48m^3$
   (Rigorous) (Skill 1.6)

   A) (1 + 2m)
   B) (1 - 8m)
   C) (1 + m - 2m)
   D) (1 - m + 2m)

5. Evaluate $3^{1/2}(9^{1/3})$
   (Rigorous) (Skill 1.7)

   A) $27^{5/6}$
   B) $9^{7/12}$
   C) $3^{5/6}$
   D) $3^{6/7}$

6. Simplify: $\sqrt{27} + \sqrt{75}$
   (Rigorous) (Skill 1.8)

   A) $8\sqrt{3}$
   B) 34
   C) $34\sqrt{3}$
   D) $15\sqrt{3}$

Mathematics 6-12

7. Which graph represents the equation of $y = x^2 + 3x$?
   (Average Rigor) (Skill 1.11)

   A)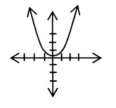
   B)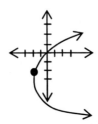

   C)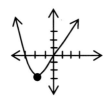
   D)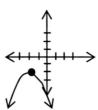

8. The volume of water flowing through a pipe varies directly with the square of the radius of the pipe. If the water flows at a rate of 80 liters per minute through a pipe with a radius of 4 cm, at what rate would water flow through a pipe with a radius of 3 cm?
   (Rigorous) (Skill 1.16)

   A) 45 liters per minute
   B) 6.67 liters per minute
   C) 60 liters per minute
   D) 4.5 liters per minute

9. What would be the shortest method of solution for the system of equations below?
   (Easy) (Skill 1.17)

   $3x + 2y = 38$
   $4x + 8 = y$

   A) linear combination
   B) additive inverse
   C) substitution
   D) graphing

10. Solve the system of equations for x, y and z.
    (Rigorous) (Skill 1.17)

    $3x + 2y - z = 0$
    $2x + 5y = 8z$
    $x + 3y + 2z = 7$

    A) $(-1, 2, 1)$
    B) $(1, 2, -1)$
    C) $(-3, 4, -1)$
    D) $(0, 1, 2)$

11. Solve for x: $18 = 4 + |2x|$
    (Rigorous) (Skill 1.20)

    A) $\{-11, 7\}$
    B) $\{-7, 0, 7\}$
    C) $\{-7, 7\}$
    D) $\{-11, 11\}$

12. Which of the following is incorrect?
    (Rigorous) (Skill 1.21)

    A) $(x^2y^3)^2 = x^4y^6$
    B) $m^2(2n)^3 = 8m^2n^3$
    C) $(m^3n^4)/(m^2n^2) = mn^2$
    D) $(x+y^2)^2 = x^2 + y^4$

13. What would be the seventh term of the expanded binomial $(2a+b)^8$?
    (Rigorous) (Skill 1.22)

    A) $2ab^7$
    B) $41a^4b^4$
    C) $112a^2b^6$
    D) $16ab^7$

14. Given a vector with horizontal component 5 and vertical component 6, determine the length of the vector.
    (Average Rigor) (Skill 1.24)

    A) 61
    B) $\sqrt{61}$
    C) 30
    D) $\sqrt{30}$

15. State the domain of the function $f(x) = \dfrac{3x-6}{x^2-25}$
    (Average Rigor) (Skill 2.3)

    A) $x \neq 2$
    B) $x \neq 5, -5$
    C) $x \neq 2, -2$
    D) $x \neq 5$

16. Find the zeroes of $f(x) = x^3 + x^2 - 14x - 24$
    (Rigorous) (Skill 2.6)

    A) 4, 3, 2
    B) 3, -8
    C) 7, -2, -1
    D) 4, -3, -2

17. $f(x) = 3x - 2;\ f^{-1}(x) =$
    (Average Rigor) (Skill 2.8)

    A) $3x + 2$
    B) $x/6$
    C) $2x - 3$
    D) $(x+2)/3$

18. Given $f(x) = 3x - 2$ and $g(x) = x^2$, determine $g(f(x))$.
    (Average Rigor) (Skill 2.9)

    A) $3x^2 - 2$
    B) $9x^2 + 4$
    C) $9x^2 - 12x + 4$
    D) $3x^3 - 2$

19. The mass of a Chips Ahoy cookie would be to
    (Average Rigor) (Skill 3.2)

    A) 1 kilogram
    B) 1 gram
    C) 15 grams
    D) 15 milligrams

20. Which term most accurately describes two coplanar lines without any common points?
    (Easy) (Skill 3.4)

    A) perpendicular
    B) parallel
    C) intersecting
    D) skew

21. What is the degree measure of an interior angle of a regular 10 sided polygon?
   (Rigorous) (Skill 3.5)

   A) 18°
   B) 36°
   C) 144°
   D) 54°

22. Given that QO⊥NP and QO=NP, quadrilateral NOPQ can most accurately be described as a
   (Easy) (Skill 3.7)

   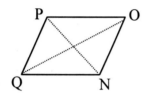

   A) parallelogram
   B) rectangle
   C) square
   D) rhombus

23. Which theorem can be used to prove $\triangle BAK \cong \triangle MKA$?
   (Average Rigor) (Skill 3.8)

   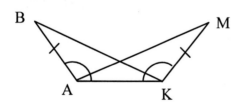

   A) SSS
   B) ASA
   C) SAS
   D) AAS

24. If a ship sails due south 6 miles, then due west 8 miles, how far was it from the starting point?
   (Average Rigor) (Skill 3.10)

   A) 100 miles
   B) 10 miles
   C) 14 miles
   D) 48 miles

25. Compute the area of the shaded region, given a radius of 5 meters. 0 is the center.
   (Rigorous) (Skill 3.12)

   A) 7.13 cm²
   B) 7.13 m²
   C) 78.5 m²
   D) 19.63 m²

   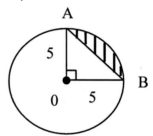

26. Determine the area of the shaded region of the trapezoid in terms of x and y.
   (Rigorous) (Skill 3.12)

   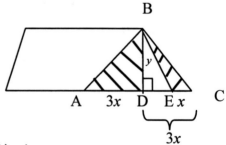

   A) $4xy$
   B) $2xy$
   C) $3x^2 y$
   D) There is not enough information given.

27. Given a 30 meter x 60 meter garden with a circular fountain with a 5 meter radius, calculate the area of the portion of the garden not occupied by the fountain.
    (Average Rigor) (Skill 3.12)

    A) 1721 m²
    B) 1879 m²
    C) 2585 m²
    D) 1015 m²

28. What is the measure of minor arc AD, given measure of arc PS is 40° and $m < K = 10°$?
    (Rigorous) (Skill 3.14)

    A) 50°
    B) 20°
    C) 30°
    D) 25°

29. Choose the diagram which illustrates the construction of a perpendicular to the line at a given point on the line.
    (Rigorous) (Skill 3.15)

    A)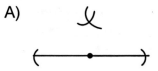

    B)

    C)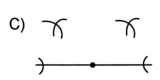

    D)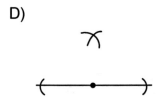

30. If the area of the base of a cone is tripled, the volume will be
    (Rigorous) (Skill 3.19)

    A) the same as the original
    B) 9 times the original
    C) 3 times the original
    D) 3 π times the original

31. Find the surface area of a box which is 3 feet wide, 5 feet tall, and 4 feet deep.
    (Easy) (Skill 3.19)

    A) 47 sq. ft.
    B) 60 sq. ft.
    C) 94 sq. ft
    D) 188 sq. ft.

32. Compute the distance from (-2,7) to the line x = 5.
    (Average Rigor) (Skill 4.1)

    A) -9
    B) -7
    C) 5
    D) 7

33. Given $K(-4, y)$ and $M(2,-3)$ with midpoint $L(x,1)$, determine the values of x and y.
    (Rigorous) (Skill 4.1)

    A) $x = -1, y = 5$
    B) $x = 3, y = 2$
    C) $x = 5, y = -1$
    D) $x = -1, y = -1$

34. Find the length of the major axis of $x^2 + 9y^2 = 36$.
    (Rigorous) (Skill 4.2)

    A) 4
    B) 6
    C) 12
    D) 8

35. Which equation represents a circle with a diameter whose endpoints are $(0,7)$ and $(0,3)$?
    (Rigorous) (Skill 4.4)

    A) $x^2 + y^2 + 21 = 0$
    B) $x^2 + y^2 - 10y + 21 = 0$
    C) $x^2 + y^2 - 10y + 9 = 0$
    D) $x^2 - y^2 - 10y + 9 = 0$

36. Which expression is equivalent to $1 - \sin^2 x$?
    (Rigorous) (Skill 5.2)

    A) $1 - \cos^2 x$
    B) $1 + \cos^2 x$
    C) $1/\sec x$
    D) $1/\sec^2 x$

37. Which expression is not equal to sinx?
    (Average Rigor) (Skill 5.2)

    A) $\sqrt{1 - \cos^2 x}$
    B) $\tan x \cos x$
    C) $1/\csc x$
    D) $1/\sec x$

38. Determine the measures of angles A and B.
(Average Rigor) (Skill 5.5)

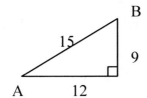

A) A = 30°, B = 60°
B) A = 60°, B = 30°
C) A = 53°, B = 37°
D) A = 37°, B = 53°

39. Compute the median for the following data set:
(Easy) (Skill 6.2)

{12, 19, 13, 16, 17, 14}

A) 14.5
B) 15.17
C) 15
D) 16

40. Half the students in a class scored 80% on an exam, most of the rest scored 85% except for one student who scored 10%. Which would be the best measure of central tendency for the test scores?
(Rigorous) (Skill 6.3)

A) mean
B) median
C) mode
D) either the median or the mode because they are equal

41. Compute the standard deviation for the following set of temperatures.
(37, 38, 35, 37, 38, 40, 36, 39)
(Easy) (Skill 6.4)

A) 37.5
B) 1.5
C) 0.5
D) 2.5

42. What conclusion can be drawn from the graph below?

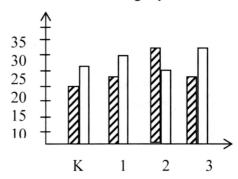

MLK Elementary
Student Enrollment   Girls   Boys
(Easy) (Skill 6.5)

A) The number of students in first grade exceeds the number in second grade.
B) There are more boys than girls in the entire school.
C) There are more girls than boys in the first grade.
D) Third grade has the largest number of students.

43. If there are three people in a room, what is the probability that at least two of them will share a birthday? (Assume a year has 365 days)
    (Rigorous) (Skill 7.1)

    A) 0.67
    B) 0.05
    C) 0.008
    D) 0.33

44. A jar contains 3 red marbles, 5 white marbles, 1 green marble and 15 blue marbles. If one marble is picked at random from the jar, what are the odds that it will be red?
    (Easy) (Skill 7.2)

    A) 1/3
    B) 1/8
    C) 3/8
    D) 1/24

45. How many ways are there to choose a potato and two green vegetables from a choice of three potatoes and seven green vegetables?
    (Average Rigor) (Skill 7.5)

    A) 126
    B) 63
    C) 21
    D) 252

46. Find the sum of the first one hundred terms in the progression.
    (-6, -2, 2 . . .)
    (Rigorous) (Skill 8.3)

    A) 19,200
    B) 19,400
    C) -604
    D) 604

47. What is the sum of the first 20 terms of the geometric sequence (2,4,8,16,32,...)?
    (Average Rigor) (Skill 8.3)

    A) 2097150
    B) 1048575
    C) 524288
    D) 1048576

48. Determine the number of subsets of set K.
    K = {4, 5, 6, 7}
    (Average Rigor) (Skill 8.4)

    A) 15
    B) 16
    C) 17
    D) 18

49. Find the value of the determinant of the matrix.
    (Average Rigor) (Skill 8.5)

    $$\begin{vmatrix} 2 & 1 & -1 \\ 4 & -1 & 4 \\ 0 & -3 & 2 \end{vmatrix}$$

    A) 0
    B) 23
    C) 24
    D) 40

50. Rewrite the following system of equations as a matrix equation.
    (Easy) (Skill 8.6)

    $3x - z = 4$
    $-x + 4y + 8z = -7$
    $x + y = z$

    A) $\begin{pmatrix} 3 & 0 & -1 \\ -1 & 4 & 8 \\ 1 & 1 & -1 \end{pmatrix} \cdot \begin{pmatrix} 4 \\ -7 \\ 0 \end{pmatrix} = \begin{pmatrix} x \\ y \\ z \end{pmatrix}$

    B) $\begin{pmatrix} 3 & 0 & -1 \\ -1 & 4 & 8 \\ 1 & 1 & -1 \end{pmatrix} \cdot \begin{pmatrix} x \\ y \\ -1 \end{pmatrix} = \begin{pmatrix} 4 \\ -7 \\ z \end{pmatrix}$

    C) $\begin{pmatrix} 3 & 0 & -1 \\ -1 & 4 & 8 \\ 1 & 1 & -1 \end{pmatrix} \cdot \begin{pmatrix} x \\ y \\ z \end{pmatrix} = \begin{pmatrix} 4 \\ -7 \\ 0 \end{pmatrix}$

    D) $\begin{pmatrix} 3 & 0 & 1 \\ 1 & 4 & 8 \\ 1 & 1 & 1 \end{pmatrix} \cdot \begin{pmatrix} x \\ y \\ z \end{pmatrix} = \begin{pmatrix} 4 \\ 7 \\ 0 \end{pmatrix}$

51. Find the first derivative of the function:
    $f(x) = x^3 - 6x^2 + 5x + 4$
    (Rigorous) (Skill 9.2)

    A) $3x^3 - 12x^2 + 5x = f'(x)$
    B) $3x^2 - 12x - 5 = f'(x)$
    C) $3x^2 - 12x + 9 = f'(x)$
    D) $3x^2 - 12x + 5 = f'(x)$

52. Differentiate: $y = e^{3x+2}$
    (Rigorous) (Skill 9.2)

    A) $3e^{3x+2} = y'$
    B) $3e^{3x} = y'$
    C) $6e^3 = y'$
    D) $(3x+2)e^{3x+1} = y'$

53. Find the slope of the line tangent to $y = 3x(\cos x)$ at $(\pi/2, \pi/2)$.
    (Rigorous) (Skill 9.6)

    A) $-3\pi/2$
    B) $3\pi/2$
    C) $\pi/2$
    D) $-\pi/2$

54. Find the equation of the line tangent to $y = 3x^2 - 5x$ at $(1, -2)$.
    (Rigorous) (Skill 9.6)

    A) $y = x - 3$
    B) $y = 1$
    C) $y = x + 2$
    D) $y = x$

55. How does the function $y = x^3 + x^2 + 4$ behave from $x = 1$ to $x = 3$?
    (Average Rigor) (Skill 9.7)

    A) increasing, then decreasing
    B) increasing
    C) decreasing
    D) neither increasing nor decreasing

56. Find the absolute maximum obtained by the function $y = 2x^2 + 3x$ on the interval $x = 0$ to $x = 3$.
    (Rigorous) (Skill 9.8)

    A) −3/4
    B) −4/3
    C) 0
    D) 27

57. The acceleration of a particle is dv/dt = 6 m/s². Find the velocity at t=10 given an initial velocity of 15 m/s.
    (Average Rigor) (Skill 9.11)

    A) 60 m/s
    B) 150 m/s
    C) 75 m/s
    D) 90 m/s

58. If the velocity of a body is given by v = 16 - t², find the distance traveled from t = 0 until the body comes to a complete stop.
    (Average Rigor) (Skill 9.11)

    A) 16
    B) 43
    C) 48
    D) 64

59. Find the antiderivative for $4x^3 - 2x + 6 = y$.
    (Rigorous) (Skill 9.13)

    A) $x^4 - x^2 + 6x + C$
    B) $x^4 - 2/3x^3 + 6x + C$
    C) $12x^2 - 2 + C$
    D) $4/3x^4 - x^2 + 6x + C$

60. Find the antiderivative for the function $y = e^{3x}$.
    (Rigorous) (Skill 9.13)

    A) $3x(e^{3x}) + C$
    B) $3(e^{3x}) + C$
    C) $1/3(e^x) + C$
    D) $1/3(e^{3x}) + C$

61. Evaluate: $\int (x^3 + 4x - 5)dx$
    (Rigorous) (Skill 9.13)

    A) $3x^2 + 4 + C$
    B) $\frac{1}{4}x^4 - 2/3x^3 + 6x + C$
    C) $x^{4/3} + 4x - 5x + C$
    D) $x^3 + 4x^2 - 5x + C$

62. Find the area under the function $y = x^2 + 4$ from $x = 3$ to $x = 6$.
    (Average Rigor) (Skill 9.14)

    A) 75
    B) 21
    C) 96
    D) 57

63. Evaluate $\int_0^2 (x^2 + x - 1)dx$
    (Rigorous) (Skill 9.15)

    A) 11/3
    B) 8/3
    C) -8/3
    D) -11/3

64. Which of the following sets is closed under division?
    (Average Rigor) (Skill 10.1)

    I) {½, 1, 2, 4}
    II) {-1, 1}
    III) {-1, 0, 1}

    A) I only
    B) II only
    C) III only
    D) I and II

65. Which of the following illustrates an inverse property?
    (Easy) (Skill 10.1)

    A) a + b = a - b
    B) a + b = b + a
    C) a + 0 = a
    D) a + (-a) = 0

66. Simplify: $\dfrac{10}{1+3i}$
    (Average Rigor) (Skill 10.2)

    A) $-1.25(1-3i)$
    B) $1.25(1+3i)$
    C) $1+3i$
    D) $1-3i$

67. Which equation corresponds to the logarithmic statement: $\log_x k = m$?
    (Rigorous) (Skill 10.3)

    A) $x^m = k$
    B) $k^m = x$
    C) $x^k = m$
    D) $m^x = k$

68. Which of the following is always composite if x is odd, y is even, and both x and y are greater than or equal to 2?
    (Average Rigor) (Skill 10.4)

    A) $x+y$
    B) $3x+2y$
    C) $5xy$
    D) $5x+3y$

69. Express .0000456 in scientific notation.
    (Easy) (Skill 10.5)

    A) $4.56x10^{-4}$
    B) $45.6x10^{-6}$
    C) $4.56x10^{-6}$
    D) $4.56x10^{-5}$

70. A student had 60 days to appeal the results of an exam. If the results were received on March 23, what was the last day that the student could appeal?
    (Average Rigor) (Skill 11.2)

    A) May 21
    B) May 22
    C) May 23
    D) May 24

71. Given the series of examples below, what is 5¢4?
    (Average Rigor) (Skill 11.3)

    4¢3=13   7¢2=47
    3¢1=8   1¢5=-4

    A) 20
    B) 29
    C) 1
    D) 21

Mathematics 6-12

72. Which of the following is a valid argument?
    (Average Rigor) (Skill 12.3)

    A) Given: if p then q; q.
       Therefore: p.
    B) Given: if p then q; ~p.
       Therefore: ~q.
    C) Given: if p then q; p.
       Therefore: q.
    D) Given: if p then q; ~q.
       Therefore: p.

73. When you begin by assuming the conclusion of a theorem is false, then show that through a sequence of logically correct steps you contradict an accepted fact, this is known as
    (Easy) (Skill 12.4)

    A) inductive reasoning
    B) direct proof
    C) indirect proof
    D) exhaustive proof

74. Which of the following best describes the process of induction?
    (Average Rigor) (Skill 12.5)

    A) making an inference based on a set of universal laws
    B) making an inference based on conjecture
    C) making an inference based on a set of concrete examples
    D) making an inference based on mathematical principles

75. Which statement most accurately describes the concept of a power?
    (Easy) (Skill 13.2)

    A) A power is repeated addition.
    B) A power is repeated integration.
    C) A power is repeated differentiation.
    D) A power is repeated multiplication.

76. What would be the least appropriate use for handheld calculators in the classroom?
    (Average Rigor) (Skill 14.1)

    A) practice for standardized tests
    B) integrating algebra and geometry with applications
    C) justifying statements in geometric proofs
    D) applying the law of sines to find dimensions

77. Which of the following is the best example of the value of personal computers in advanced high school mathematics?
(Easy) (Skill 14.1)

   A) Students can independently drill and practice test questions.
   B) Students can keep an organized list of theorems and postulates on a word processing program.
   C) Students can graph and calculate complex functions to explore their nature and make conjectures.
   D) Students are better prepared for business because of mathematics computer programs in high school.

78. Identify the correct sequence of subskills required for solving and graphing inequalities involving absolute value in one variable, such as $|x+1| \leq 6$.
(Average Rigor) (Skill 14.2)

   A) understanding absolute value, graphing inequalities, solving systems of equations
   B) graphing inequalities on a Cartesian plane, solving systems of equations, simplifying expressions with absolute value
   C) plotting points, graphing equations, graphing inequalities
   D) solving equations with absolute value, solving inequalities, graphing conjunctions and disjunctions

79. A group of students working with trigonometric identities have concluded that $\cos 2x = 2\cos x$. How could you best lead them to discover their error?
(Average Rigor) (Skill 14.2)

   A) Have the students plug in values on their calculators.
   B) Direct the student to the appropriate chapter in the text.
   C) Derive the correct identity on the board.
   D) Provide each student with a table of trig identities.

80. $-3 + 7 = -4$ $6(-10) = -60$
    $-5(-15) = 75$ $-3 + -8 = 11$
    $8 - 12 = -4$ $7 - -8 = 15$

    Which best describes the type of error observed above?
    (Easy) (Skill 15.1)

    A) The student is incorrectly multiplying integers.
    B) The student has incorrectly applied rules for adding integers to subtracting integers.
    C) The student has incorrectly applied rules for multiplying integers to adding integers.
    D) The student is incorrectly subtracting integers.

ANSWER KEY

1) D	17) D	34) C	51) D	68) C
2) B	18) C	35) B	52) A	69) D
3) D	19) C	36) D	53) A	70) B
4) A	20) B	37) D	54) A	71) D
5) B	21) C	38) D	55) B	72) C
6) A	22) C	39) C	56) D	73) C
7) C	23) C	40) B	57) C	74) C
8) A	24) B	41) B	58) B	75) D
9) C	25) B	42) B	59) A	76) C
10) A	26) B	43) C	60) D	77) C
11) C	27) A	44) B	61) B	78) D
12) D	28) B	45) A	62) A	79) C
13) C	29) D	46) A	63) B	80) C
14) B	30) C	47) A	64) B	
15) B	31) C	48) B	65) D	
16) D	32) D	49) C	66) D	
	33) A	50) C	67) A	

Rigor Table

	Easy %20	Average Rigor %40	Rigorous %40
Question #	2, 9, 20, 22, 31, 39, 41, 42, 44, 50, 65, 69, 73, 75, 77, 80	1, 3, 7, 14, 15, 17, 18, 19, 23, 24, 27, 32, 37, 38, 45, 47, 48, 49, 55, 57, 58, 62, 64, 66, 68, 70, 71, 72, 74, 76, 78, 79	4, 5, 6, 8, 10, 11, 12, 13, 16, 21, 25, 26, 28, 29, 30, 33, 34, 35, 36, 40, 43, 46, 51, 52, 53, 54, 56, 59, 60, 61, 63, 67

Rationales with Sample Questions

1. Which graph represents the solution set for $x^2 - 5x > -6$?
   (Average Rigor) (Skill 1.1)

   A) number line with open circles at -2 and 2

   B) number line with open circles at -3 and 3

   C) number line with open circles at -2 and 2

   D) number line with open circles at 2 and 3, shaded outside

Answer: D

Rewriting the inequality gives $x^2 - 5x + 6 > 0$. Factoring gives $(x - 2)(x - 3) > 0$. The two cut-off points on the number line are now at $x = 2$ and $x = 3$. Choosing a random number in each of the three parts of the numberline, we test them to see if they produce a true statement. If $x = 0$ or $x = 4$, $(x-2)(x-3)>0$ is true. If $x = 2.5$, $(x-2)(x-3)>0$ is false. Therefore the solution set is all numbers smaller than 2 or greater than 3.

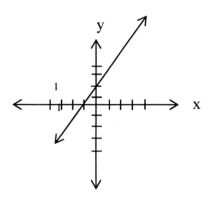

2. What is the equation of the above graph?
   (Easy) (Skill 1.2)

A) $2x + y = 2$
B) $2x - y = -2$
C) $2x - y = 2$
D) $2x + y = -2$

Answer: B

By observation, we see that the graph has a y-intercept of 2 and a slope of 2/1 = 2. Therefore its equation is y = mx + b = 2x + 2. Rearranging the terms gives 2x − y = -2.

3. Solve for v_0 : $d = at(v_t - v_0)$
   (Average Rigor) (Skill 1.5)

A) $v_0 = atd - v_t$
B) $v_0 = d - atv_t$
C) $v_0 = atv_t - d$
D) $v_0 = (atv_t - d)/at$

Answer: D

Using the Distributive Property and other properties of equality to isolate v_0 gives
$d = atv_t - atv_0$, $atv_0 = atv_t - d$, $v_0 = \dfrac{atv_t - d}{at}$.

TEACHER CERTIFICATION STUDY GUIDE

4. **Which of the following is a factor of** $6+48m^3$
   (Rigorous) (Skill 1.6)

   A) (1 + 2m)
   B) (1 - 8m)
   C) (1 + m - 2m)
   D) (1 - m + 2m)

   Answer: A

   Removing the common factor of 6 and then factoring the sum of two cubes gives
   $6 + 48m^3 = 6(1 + 8m^3) = 6(1 + 2m)(1^2 − 2m + (2m)^2)$.

5. **Evaluate** $3^{1/2}(9^{1/3})$
   (Rigorous) (Skill 1.7)

   A) $27^{5/6}$
   B) $9^{7/12}$
   C) $3^{5/6}$
   D) $3^{6/7}$

   Answer: B

   Getting the bases the same gives us $3^{\frac{1}{2}}3^{\frac{2}{3}}$. Adding exponents gives $3^{\frac{7}{6}}$. Then some additional manipulation of exponents produces $3^{\frac{7}{6}} = 3^{\frac{14}{12}} = (3^2)^{\frac{7}{12}} = 9^{\frac{7}{12}}$.

6. **Simplify:** $\sqrt{27} + \sqrt{75}$
   (Rigorous) (Skill 1.8)

   A) $8\sqrt{3}$
   B) 34
   C) $34\sqrt{3}$
   D) $15\sqrt{3}$

   Answer: A

   Simplifying radicals gives $\sqrt{27} + \sqrt{75} = 3\sqrt{3} + 5\sqrt{3} = 8\sqrt{3}$.

7. Which graph represents the equation of $y = x^2 + 3x$?
   (Average Rigor) (Skill 1.11)

A)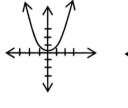

B)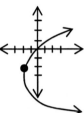

C)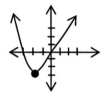

D)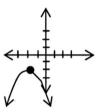

Answer: C

B is not the graph of a function. D is the graph of a parabola where the coefficient of x^2 is negative. A appears to be the graph of $y = x^2$. To find the x-intercepts of $y = x^2 + 3x$, set $y = 0$ and solve for x: $0 = x^2 + 3x = x(x + 3)$ to get $x = 0$ or $x = -3$. Therefore, the graph of the function intersects the x-axis at x=0 and x=-3.

8. The volume of water flowing through a pipe varies directly with the square of the radius of the pipe. If the water flows at a rate of 80 liters per minute through a pipe with a radius of 4 cm, at what rate would water flow through a pipe with a radius of 3 cm?
(Rigorous) (Skill 1.16)

A) 45 liters per minute
B) 6.67 liters per minute
C) 60 liters per minute
D) 4.5 liters per minute

Answer: A

Set up the direct variation: $\frac{V}{r^2} = \frac{V}{r^2}$. Substituting gives $\frac{80}{16} = \frac{V}{9}$. Solving for V gives 45 liters per minute.

9. What would be the shortest method of solution for the system of equations below?
(Easy) (Skill 1.17)

$$3x + 2y = 38$$
$$4x + 8 = y$$

A) linear combination
B) additive inverse
C) substitution
D) graphing

Answer: C

Since the second equation is already solved for y, it would be easiest to use the substitution method.

TEACHER CERTIFICATION STUDY GUIDE

10. **Solve the system of equations** for x, y and z.
    (Rigorous) (Skill 1.17)

    $3x + 2y - z = 0$
    $2x + 5y = 8z$
    $x + 3y + 2z = 7$

    A) $(-1, 2, 1)$
    B) $(1, 2, -1)$
    C) $(-3, 4, -1)$
    D) $(0, 1, 2)$

    Answer: A

    Multiplying equation 1 by 2, and equation 2 by –3, and then adding together the two resulting equations gives -11y + 22z = 0. Solving for y gives y = 2z. In the meantime, multiplying equation 3 by –2 and adding it to equation 2 gives –y – 12z = -14. Then substituting 2z for y, yields the result z = 1. Subsequently, one can easily find that y = 2, and x = -1.

11. **Solve for** x: $18 = 4 + |2x|$
    (Rigorous) (Skill 1.20)

    A) $\{-11, 7\}$
    B) $\{-7, 0, 7\}$
    C) $\{-7, 7\}$
    D) $\{-11, 11\}$

    Answer: C

    Using the definition of absolute value, two equations are possible: 18 = 4 + 2x or 18 = 4 – 2x. Solving for x gives x = 7 or x = -7.

12. Which of the following is incorrect?
    (Rigorous) (Skill 1.21)

A) $(x^2 y^3)^2 = x^4 y^6$
B) $m^2 (2n)^3 = 8m^2 n^3$
C) $(m^3 n^4)/(m^2 n^2) = mn^2$
D) $(x+y^2)^2 = x^2 + y^4$

Answer: D

Using FOIL to do the expansion, we get $(x + y^2)^2 = (x + y^2)(x + y^2) = x^2 + 2xy^2 + y^4$.

13. What would be the seventh term of the expanded binomial $(2a+b)^8$?
    (Rigorous) (Skill 1.22)

A) $2ab^7$
B) $41a^4 b^4$
C) $112a^2 b^6$
D) $16ab^7$

Answer: C

The set-up for finding the seventh term is $\frac{8(7)(6)(5)(4)(3)}{6(5)(4)(3)(2)(1)}(2a)^{8-6} b^6$ which gives $28(4a^2 b^6)$ or $112a^2 b^6$.

14. Given a vector with horizontal component 5 and vertical component 6, determine the length of the vector.
    (Average Rigor) (Skill 1.24)

A) 61
B) $\sqrt{61}$
C) 30
D) $\sqrt{30}$

Answer: B

Using the Pythagorean Theorem, we get v = $\sqrt{36 + 25} = \sqrt{61}$.

15. **State the domain of the function** $f(x) = \dfrac{3x-6}{x^2-25}$

    (Average Rigor) (Skill 2.3)

    A) $x \neq 2$
    B) $x \neq 5, -5$
    C) $x \neq 2, -2$
    D) $x \neq 5$

    Answer: B

    The values of 5 and –5 must be omitted from the domain of all real numbers because if x took on either of those values, the denominator of the fraction would have a value of 0, and therefore the fraction would be undefined.

16. **Find the zeroes of** $f(x) = x^3 + x^2 - 14x - 24$

    (Rigorous) (Skill 2.6)

    A) 4, 3, 2
    B) 3, -8
    C) 7, -2, -1
    D) 4, -3, -2

    Answer: D

    Possible rational roots of the equation 0 = x³ + x² – 14x -24 are all the positive and negative factors of 24. By substituting into the equation, we find that –2 is a root, and therefore that x+2 is a factor. By performing the long division (x³ + x² – 14x – 24)/(x+2), we can find that another factor of the original equation is x² – x – 12 or (x-4)(x+3). Therefore the zeros of the original function are –2, -3, and 4.

17. $f(x) = 3x - 2;\ f^{-1}(x) =$
(Average Rigor) (Skill 2.8)

A) $3x + 2$
B) $x/6$
C) $2x - 3$
D) $(x+2)/3$

Answer: D

To find the inverse, f⁻¹(x), of the given function, reverse the variables in the given equation, y = 3x – 2, to get x = 3y – 2. Then solve for y as follows:
x+2 = 3y, and y = $\frac{x+2}{3}$.

18. Given $f(x) = 3x - 2$ and $g(x) = x^2$, determine $g(f(x))$.
(Average Rigor) (Skill 2.9)

A) $3x^2 - 2$
B) $9x^2 + 4$
C) $9x^2 - 12x + 4$
D) $3x^3 - 2$

Answer: C

The composite function g(f(x)) = (3x-2)² = 9x² – 12x + 4.

19. The mass of a Chips Ahoy cookie would be to
(Average Rigor) (Skill 3.2)

A) 1 kilogram
B) 1 gram
C) 15 grams
D) 15 milligrams

Answer: C

Since an ordinary cookie would not weigh as much as 1 kilogram, or as little as 1 gram or 15 milligrams, the only reasonable answer is 15 grams.

20. Which term most accurately describes two coplanar lines without any common points?
    (Easy) (Skill 3.4)

    A) perpendicular
    B) parallel
    C) intersecting
    D) skew

    Answer: B

    By definition, parallel lines are coplanar lines without any common points.

21. What is the degree measure of an interior angle of a regular 10 sided polygon?
    (Rigorous) (Skill 3.5)

    A) 18°
    B) 36°
    C) 144°
    D) 54°

    Answer: C

    Formula for finding the measure of each interior angle of a regular polygon with n sides is $\frac{(n-2)180}{n}$. For n=10, we get $\frac{8(180)}{10} = 144$.

22. Given that QO⊥NP and QO=NP, quadrilateral NOPQ can most accurately be described as a
 (Easy) (Skill 3.6)

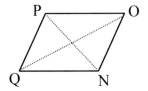

A) parallelogram
B) rectangle
C) square
D) rhombus

Answer: C

In an ordinary parallelogram, the diagonals are not perpendicular or equal in length. In a rectangle, the diagonals are not necessarily perpendicular. In a rhombus, the diagonals are not equal in length. In a square, the diagonals are both perpendicular and congruent.

23. Which theorem can be used to prove $\triangle BAK \cong \triangle MKA$?
 (Average Rigor) (Skill 3.8)

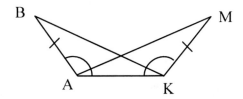

A) SSS
B) ASA
C) SAS
D) AAS

Answer: C

Since side AK is common to both triangles, the triangles can be proved congruent by using the Side-Angle-Side Postulate.

24. **If a ship sails due south 6 miles, then due west 8 miles, how far was it from the starting point?**
    (Average Rigor) (Skill 3.10)

    A) 100 miles
    B) 10 miles
    C) 14 miles
    D) 48 miles

    Answer: B

    Draw a right triangle with legs of 6 and 8. Find the hypotenuse using the Pythagorean Theorem. $6^2 + 8^2 = c^2$. Therefore, c = 10 miles.

25. **Compute the area of the shaded region, given a radius of 5 meters. 0 is the center.**
    (Rigorous) (Skill 3.12)

    A) 7.13 cm²
    B) 7.13 m²
    C) 78.5 m²
    D) 19.63 m²

    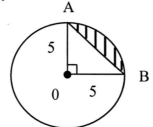

    Answer: B

    Area of triangle AOB is .5(5)(5) = 12.5 square meters. Since $\frac{90}{360} = .25$, the area of sector AOB (pie-shaped piece) is approximately $.25(\pi)5^2$ = 19.63. Subtracting the triangle area from the sector area to get the area of segment AB, we get approximately 19.63-12.5 = 7.13 square meters.

TEACHER CERTIFICATION STUDY GUIDE

26. Determine the area of the shaded region of the trapezoid in terms of *x* and *y*.
    (Rigorous) (Skill 3.12)

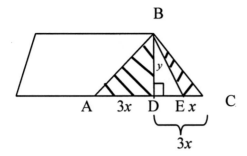

A) $4xy$
B) $2xy$
C) $3x^2y$
D) There is not enough information given.

Answer: B

To find the area of the shaded region, find the area of triangle ABC and then subtract the area of triangle DBE. The area of triangle ABC is .5(6x)(y) = 3xy. The area of triangle DBE is .5(2x)(y) = xy. The difference is 2xy.

27. Given a 30 meter x 60 meter garden with a circular fountain with a 5 meter radius, calculate the area of the portion of the garden not occupied by the fountain.
    (Average Rigor) (Skill 3.12)

A) 1721 m²
B) 1879 m²
C) 2585 m²
D) 1015 m²

Answer: A

Find the area of the garden and then subtract the area of the fountain: $30(60) - \pi(5)^2$ or approximately 1721 square meters.

28. What is the measure of minor arc AD, given measure of arc PS is 40° and $m\angle K = 10°$?
   (Rigorous) (Skill 3.14)

   A) 50°
   B) 20°
   C) 30°
   D) 25°

   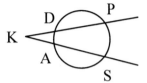

Answer: B

The formula relating the measure of angle K and the two arcs it intercepts is $m\angle K = \frac{1}{2}(mPS - mAD)$. Substituting the known values, we get $10 = \frac{1}{2}(40 - mAD)$. Solving for mAD gives an answer of 20 degrees.

29. Choose the diagram which illustrates the construction of a perpendicular to the line at a given point on the line.
   (Rigorous) (Skill 3.15)

A)

B)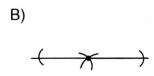

C)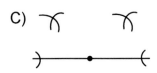

D)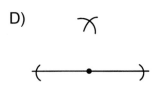

Answer: D

Given a point on a line, place the compass point there and draw two arcs intersecting the line in two points, one on either side of the given point. Then using any radius larger than half the new segment produced, and with the pointer at each end of the new segment, draw arcs which intersect above the line. Connect this new point with the given point.

30. **If the area of the base of a cone is tripled, the volume will be**
    (Rigorous) (Skill 3.19)

    A) the same as the original
    B) 9 times the original
    C) 3 times the original
    D) 3π times the original

Answer: C

The formula for the volume of a cone is $V = \frac{1}{3}Bh$, where B is the area of the circular base and h is the height. If the area of the base is tripled, the volume becomes $V = \frac{1}{3}(3B)h = Bh$, or three times the original area.

31. **Find the surface area of a box which is 3 feet wide, 5 feet tall, and 4 feet deep.**
    (Easy) (Skill 3.19)

    A) 47 sq. ft.
    B) 60 sq. ft.
    C) 94 sq. ft
    D) 188 sq. ft.

Answer: C

Let's assume the base of the rectangular solid (box) is 3 by 4, and the height is 5. Then the surface area of the top and bottom together is 2(12) = 24. The sum of the areas of the front and back are 2(15) = 30, while the sum of the areas of the sides are 2(20)=40. The total surface area is therefore 94 square feet.

TEACHER CERTIFICATION STUDY GUIDE

32. Compute the distance from (-2,7) to the line x = 5.
    (Average Rigor) (Skill 4.1)

A) -9
B) -7
C) 5
D) 7

Answer: D

The line x = 5 is a vertical line passing through (5,0) on the Cartesian plane. By observation the distance along the horizontal line from the point (-2,7) to the line x=5 is 7 units.

33. Given $K(-4, y)$ and $M(2,-3)$ with midpoint $L(x,1)$, determine the values of x and y.
    (Rigorous) (Skill 4.1)

A) $x = -1, y = 5$
B) $x = 3, y = 2$
C) $x = 5, y = -1$
D) $x = -1, y = -1$

Answer: A

The formula for finding the midpoint (a,b) of a segment passing through the points (x_1, y_1) and (x_2, y_2) is $(a,b) = (\frac{x_1 + x_2}{2}, \frac{y_1 + y_2}{2})$. Setting up the corresponding equations from this information gives us $x = \frac{-4+2}{2}, and\ 1 = \frac{y-3}{2}$. Solving for x and y gives x = -1 and y = 5.

Mathematics 6-12

34. Find the length of the major axis of $x^2 + 9y^2 = 36$.

    (Rigorous) (Skill 4.2)

    A) 4
    B) 6
    C) 12
    D) 8

Answer: C

Dividing by 36, we get $\frac{x^2}{36} + \frac{y^2}{4} = 1$, which tells us that the ellipse intersects the x-axis at 6 and –6, and therefore the length of the major axis is 12. (The ellipse intersects the y-axis at 2 and –2).

35. Which equation represents a circle with a diameter whose endpoints are $(0, 7)$ and $(0, 3)$?

    (Rigorous) (Skill 4.3)

    A) $x^2 + y^2 + 21 = 0$
    B) $x^2 + y^2 - 10y + 21 = 0$
    C) $x^2 + y^2 - 10y + 9 = 0$
    D) $x^2 - y^2 - 10y + 9 = 0$

Answer: B

With a diameter going from (0,7) to (0,3), the diameter of the circle must be 4, the radius must be 2, and the center of the circle must be at (0,5). Using the standard form for the equation of a circle, we get $(x-0)^2 + (y-5)^2 = 2^2$. Expanding, we get $x^2 + y^2 - 10y + 21 = 0$.

36. Which expression is equivalent to $1-\sin^2 x$?
    (Rigorous) (Skill 5.2)

A) $1-\cos^2 x$
B) $1+\cos^2 x$
C) $1/\sec x$
D) $1/\sec^2 x$

Answer: D

Using the Pythagorean Identity, we know sin²x + cos²x = 1. Thus 1 – sin²x = cos²x, which by definition is equal to 1/sec²x.

37. Which expression is not equal to sinx?
    (Average Rigor) (Skill 5.2)

A) $\sqrt{1-\cos^2 x}$
B) $\tan x \cos x$
C) $1/\csc x$
D) $1/\sec x$

Answer: D

Using the basic definitions of the trigonometric functions and the Pythagorean identity, we see that the first three options are all identical to sinx. secx= 1/cosx is not the same as sinx.

38. **Determine the measures of angles A and B.**
    (Average Rigor) (Skill 5.5)

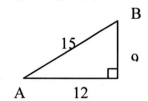

A) A = 30°, B = 60°
B) A = 60°, B = 30°
C) A = 53°, B = 37°
D) A = 37°, B = 53°

Answer: D

Tan A = 9/12=.75 and tan^{-1}.75 = 37 degrees. Since angle B is complementary to angle A, the measure of angle B is therefore 53 degrees.

39. **Compute the median for the following data set:**
    (Easy) (Skill 6.2)

    {12, 19, 13, 16, 17, 14}

A) 14.5
B) 15.17
C) 15
D) 16

Answer: C

Arrange the data in ascending order: 12,13,14,16,17,19. The median is the middle value in a list with an odd number of entries. When there is an even number of entries, the median is the mean of the two center entries. Here the average of 14 and 16 is 15.

TEACHER CERTIFICATION STUDY GUIDE

40. Half the students in a class scored 80% on an exam, most of the rest scored 85% except for one student who scored 10%. Which would be the best measure of central tendency for the test scores?
    (Rigorous) (Skill 6.3)

    A) mean
    B) median
    C) mode
    D) either the median or the mode because they are equal

    Answer: B

    In this set of data, the median (see #39) would be the most representative measure of central tendency since the median is independent of extreme values. Because of the 10% outlier, the mean (average) would be disproportionately skewed. In this data set, it is true that the median and the mode (number which occurs most often) are the same, but the median remains the best choice because of its special properties.

41. Compute the standard deviation for the following set of temperatures.
    (37, 38, 35, 37, 38, 40, 36, 39)
    (Easy) (Skill 6.4)

    A) 37.5
    B) 1.5
    C) 0.5
    D) 2.5

    Answer: B

    Find the mean: 300/8 = 37.5. Then, using the formula for standard deviation, we get

    $$\sqrt{\frac{2(37.5-37)^2 + 2(37.5-38)^2 + (37.5-35)^2 + (37.5-40)^2 + (37.5-36)^2 + (37.5-39)^2}{8}}$$

    which has a value of 1.5.

Mathematics 6-12

42. What conclusion can be drawn from the graph below?

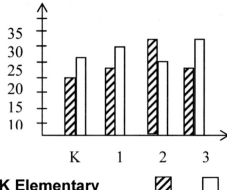

MLK Elementary Student Enrollment   Girls   Boys

(Easy) (Skill 6.5)

A) The number of students in first grade exceeds the number in second grade.
B) There are more boys than girls in the entire school.
C) There are more girls than boys in the first grade.
D) Third grade has the largest number of students.

Answer: B

In Kindergarten, first grade, and third grade, there are more boys than girls. The number of extra girls in grade two is more than made up for by the extra boys in all the other grades put together.

43. If there are three people in a room, what is the probability that at least two of them will share a birthday? (Assume a year has 365 days) *(Rigorous) (Skill 7.1)*

A) 0.67
B) 0.05
C) 0.008
D) 0.33

Answer: C

The best way to approach this problem is to use the fact that the probability of an event + the probability of the event not happening = 1. First find the probability that no two people will share a birthday and then subtract that from one. The probability that two of the people will not share a birthday = 364/365 (since the second person's birthday can be one of the 364 days other than the birthday of the first person). The probability that the third person will also not share either of the first two birthdays = (364/365) * (363/365) = 0.992. Therefore, the probability that at least two people will share a birthday = 1 – 0.992 = 0.008.

TEACHER CERTIFICATION STUDY GUIDE

44. A jar contains 3 red marbles, 5 white marbles, 1 green marble and 15 blue marbles. If one marble is picked at random from the jar, what are the odds that it will be red?
   (Easy)(Skill 7.2)

   A) 1/3
   B) 1/8
   C) 3/8
   D) 1/24

   Answer: B

   The total number of marbles is 24 and the number of red marbles is 3. Thus the odds of picking a red marble from the jar are 3/24=1/8.

45. How many ways are there to choose a potato and two green vegetables from a choice of three potatoes and seven green vegetables?
   (Average Rigor) (Skill 7.5)

   A) 126
   B) 63
   C) 21
   D) 252

   Answer: A

   There are 3 slots to fill. There are 3 choices for the first, 7 for the second, and 6 for the third. Therefore, the total number of choices is 3(7)(6) = 126.

TEACHER CERTIFICATION STUDY GUIDE

46. Find the sum of the first one hundred terms in the progression.
    (-6, -2, 2 . . .)
    (Rigorous) (Skill 8.3)

A) 19,200
B) 19,400
C) -604
D) 604

Answer: A

To find the 100th term: t_{100} = -6 + 99(4) = 390. To find the sum of the first 100 terms: $S = \frac{100}{2}(-6 + 390) = 19200$.

47. What is the sum of the first 20 terms of the geometric sequence (2,4,8,16,32,…)?
    (Average Rigor) (Skill 8.3)

A) 2097150
B) 1048575
C) 524288
D) 1048576

Answer: A

For a geometric sequence $a, ar, ar^2, ..., ar^n$, the sum of the first n terms is given by $\frac{a(r^n - 1)}{r - 1}$. In this case a=2 and r=2. Thus the sum of the first 20 terms of the sequence is given by $\frac{2(2^{20} - 1)}{2 - 1} = 2097150$.

Mathematics 6-12

48. Determine the number of subsets of set K.
   $K = \{4, 5, 6, 7\}$
   (Average Rigor) (Skill 8.4)

A) 15
B) 16
C) 17
D) 18

Answer: B

A set of n objects has 2^n subsets. Therefore, here we have $2^4 = 16$ subsets. These subsets include four which each have 1 element only, six which each have 2 elements, four which each have 3 elements, plus the original set, and the empty set.

49. Find the value of the determinant of the matrix.
(Average Rigor) (Skill 8.5)

$$\begin{vmatrix} 2 & 1 & -1 \\ 4 & -1 & 4 \\ 0 & -3 & 2 \end{vmatrix}$$

A) 0
B) 23
C) 24
D) 40

Answer: C

To find the determinant of a matrix without the use of a graphing calculator, repeat the first two columns as shown,

```
2    1    -1    2    1
4    -1    4    4    -1
0    -3    2    0    -3
```

Starting with the top left-most entry, 2, multiply the three numbers in the diagonal going down to the right: 2(-1)(2)=-4. Do the same starting with 1: 1(4)(0)=0. And starting with –1: -1(4)(-3) = 12. Adding these three numbers, we get 8. Repeat the same process starting with the top right-most entry, 1. That is, multiply the three numbers in the diagonal going down to the left: 1(4)(2) = 8. Do the same starting with 2: 2(4)(-3) = -24 and starting with –1: -1(-1)(0) = 0. Add these together to get -16. To find the determinant, subtract the second result from the first: 8-(-16)=24.

TEACHER CERTIFICATION STUDY GUIDE

50. **Rewrite the following system of equations as a matrix equation.**
    (Easy) (Skill 8.6)

$3x - z = 4$

$-x + 4y + 8z = -7$

$x + y = z$

A) $\begin{pmatrix} 3 & 0 & -1 \\ -1 & 4 & 8 \\ 1 & 1 & -1 \end{pmatrix} \cdot \begin{pmatrix} 4 \\ -7 \\ 0 \end{pmatrix} = \begin{pmatrix} x \\ y \\ z \end{pmatrix}$

B) $\begin{pmatrix} 3 & 0 & -1 \\ -1 & 4 & 8 \\ 1 & 1 & -1 \end{pmatrix} \cdot \begin{pmatrix} x \\ y \\ -1 \end{pmatrix} = \begin{pmatrix} 4 \\ -7 \\ z \end{pmatrix}$

C) $\begin{pmatrix} 3 & 0 & -1 \\ -1 & 4 & 8 \\ 1 & 1 & -1 \end{pmatrix} \cdot \begin{pmatrix} x \\ y \\ z \end{pmatrix} = \begin{pmatrix} 4 \\ -7 \\ 0 \end{pmatrix}$

D) $\begin{pmatrix} 3 & 0 & 1 \\ 1 & 4 & 8 \\ 1 & 1 & 1 \end{pmatrix} \cdot \begin{pmatrix} x \\ y \\ z \end{pmatrix} = \begin{pmatrix} 4 \\ 7 \\ 0 \end{pmatrix}$

Answer: C

It is, perhaps, easiest to write the system of equations in a full form that shows the structure more clearly. This simply involves some rearranging of the equations:

$3x + 0y - z = 4$

$-x + 4y + 8z = -7$

$x + y - z = 0$

Next, form the coefficient matrix and multiply it by the variable vector to get the solution vector:

$\begin{pmatrix} 3 & 0 & -1 \\ -1 & 4 & 8 \\ 1 & 1 & -1 \end{pmatrix} \cdot \begin{pmatrix} x \\ y \\ z \end{pmatrix} = \begin{pmatrix} 4 \\ -7 \\ 0 \end{pmatrix}.$

51. Find the first derivative of the function: $f(x) = x^3 - 6x^2 + 5x + 4$
    (Rigorous) (Skill 9.2)

    A) $3x^3 - 12x^2 + 5x = f'(x)$
    B) $3x^2 - 12x - 5 = f'(x)$
    C) $3x^2 - 12x + 9 = f'(x)$
    D) $3x^2 - 12x + 5 = f'(x)$

    Answer: D

    Use the Power Rule for polynomial differentiation: if $y = ax^n$, then $y' = nax^{n-1}$.

52. Differentiate: $y = e^{3x+2}$
    (Rigorous) (Skill 9.2)

    A) $3e^{3x+2} = y'$
    B) $3e^{3x} = y'$
    C) $6e^3 = y'$
    D) $(3x+2)e^{3x+1} = y'$

    Answer: A

    Use the Exponential Rule for derivatives of functions of e: if $y = ae^{f(x)}$, then $y' = f'(x)ae^{f(x)}$.

53. Find the slope of the line tangent to $y = 3x(\cos x)$ at $(\pi/2, \pi/2)$.
    (Rigorous) (Skill 9.6)

    A) $-3\pi/2$
    B) $3\pi/2$
    C) $\pi/2$
    D) $-\pi/2$

    Answer: A

    To find the slope of the tangent line, find the derivative, and then evaluate it at $x = \frac{\pi}{2}$. $y' = 3x(-\sin x) + 3\cos x$. At the given value of x,
    $y' = 3(\frac{\pi}{2})(-\sin\frac{\pi}{2}) + 3\cos\frac{\pi}{2} = \frac{-3\pi}{2}$.

54. **Find the equation of the line tangent to** $y = 3x^2 - 5x$ **at** $(1,-2)$.
    (Rigorous) (Skill 9.6)

    A) $y = x - 3$
    B) $y = 1$
    C) $y = x + 2$
    D) $y = x$

Answer: A

To find the slope of the tangent line, find the derivative, and then evaluate it at x=1.

y'=6x-5=6(1)-5=1. Then using point-slope form of the equation of a line, we get y+2=1(x-1) or y = x-3.

55. **How does the function** $y = x^3 + x^2 + 4$ **behave from** $x = 1$ **to** $x = 3$?
    (Average Rigor) (Skill 9.8)

    A) increasing, then decreasing
    B) increasing
    C) decreasing
    D) neither increasing nor decreasing

Answer: B

To find critical points, take the derivative, set it equal to 0, and solve for x.
f'(x) = 3x^2 + 2x = x(3x+2)=0. CP at x=0 and x=-2/3. Neither of these CP is on the interval from x=1 to x=3. Testing the endpoints: at x=1, y=6 and at x=3, y=38. Since the derivative is positive for all values of x from x=1 to x=3, the curve is increasing on the entire interval.

56. Find the absolute maximum obtained by the function $y = 2x^2 + 3x$ on the interval $x = 0$ to $x = 3$.
    (Rigorous) (Skill 9.8)

A) $-3/4$
B) $-4/3$
C) 0
D) 27

Answer: D

Find CP at x=−.75 as done in #55. Since the CP is not in the interval from x=0 to x=3, just find the values of the functions at the endpoints. When x=0, y=0, and when x=3, y = 27. Therefore 27 is the absolute maximum on the given interval.

57. The acceleration of a particle is dv/dt = 6 m/s². Find the velocity at t=10 given an initial velocity of 15 m/s.
    (Average Rigor) (Skill 9.11)

A) 60 m/s
B) 150 m/s
C) 75 m/s
D) 90 m/s

Answer: C

Recall that the derivative of the velocity function is the acceleration function. In reverse, the integral of the acceleration function is the velocity function. Therefore, if a=6, then v=6t+C. Given that at t=0, v=15, we get v = 6t+15. At t=10, v=60+15=75m/s.

58. If the velocity of a body is given by v = 16 - t², find the distance traveled from t = 0 until the body comes to a complete stop.
    (Average Rigor) (Skill 9.11)

    A) 16
    B) 43
    C) 48
    D) 64

    Answer: B

    Recall that the derivative of the distance function is the velocity function. In reverse, the integral of the velocity function is the distance function. To find the time needed for the body to come to a stop when v=0, solve for t: v = 16 – t² = 0. Result: t = 4 seconds. The distance function is s = 16t - $\frac{t^3}{3}$. At t=4, s= 64 – 64/3 or approximately 43 units.

59. Find the antiderivative for $4x^3 - 2x + 6 = y$.
    (Rigorous) (Skill 9.13)

    A) $x^4 - x^2 + 6x + C$
    B) $x^4 - 2/3x^3 + 6x + C$
    C) $12x^2 - 2 + C$
    D) $4/3x^4 - x^2 + 6x + C$

    Answer: A

    Use the rule for polynomial integration: given ax^n, the antiderivative is $\frac{ax^{n+1}}{n+1}$.

TEACHER CERTIFICATION STUDY GUIDE

60. Find the antiderivative for the function $y = e^{3x}$.
(Rigorous) (Skill 9.13)

A) $3x(e^{3x}) + C$
B) $3(e^{3x}) + C$
C) $1/3(e^x) + C$
D) $1/3(e^{3x}) + C$

Answer: D

Use the rule for integration of functions of e: $\int e^x dx = e^x + C$.

61. Evaluate: $\int (x^3 + 4x - 5) dx$
(Rigorous) (Skill 9.13)

A) $3x^2 + 4 + C$
B) $\frac{1}{4}x^4 - 2/3x^3 + 6x + C$
C) $x^{4/3} + 4x - 5x + C$
D) $x^3 + 4x^2 - 5x + C$

Answer: B

Integrate as described in #59.

62. Find the area under the function $y = x^2 + 4$ from $x = 3$ to $x = 6$.
(Average Rigor) (Skill 9.14)

A) 75
B) 21
C) 96
D) 57

Answer: A

To find the area set up the definite integral: $\int_3^6 (x^2 + 4) dx = (\frac{x^3}{3} + 4x)$. Evaluate the expression at x=6, at x=3, and then subtract to get (72+24)-(9+12)=75.

TEACHER CERTIFICATION STUDY GUIDE

63. Evaluate $\int_0^2 (x^2 + x - 1)dx$
(Rigorous) (Skill 9.15)

A) 11/3
B) 8/3
C) -8/3
D) -11/3

Answer: B

Use the fundamental theorem of calculus to find the definite integral: given a continuous function f on an interval [a,b], then $\int_a^b f(x)dx = F(b) - F(a)$, where F is an antiderivative of f.

$\int_0^2 (x^2 + x - 1)dx = (\frac{x^3}{3} + \frac{x^2}{2} - x)$ Evaluate the expression at x=2, at x=0, and then subtract to get 8/3 + 4/2 – 2-0 = 8/3.

64. Which of the following sets is closed under division?
(Average Rigor) (Skill 10.1)

    I) {½, 1, 2, 4}
    II) {-1, 1}
    III) {-1, 0, 1}

A) I only
B) II only
C) III only
D) I and II

Answer: B

I is not closed because $\frac{4}{.5} = 8$ and 8 is not in the set.

III is not closed because $\frac{1}{0}$ is undefined.

II is closed because $\frac{-1}{1} = -1, \frac{1}{-1} = -1, \frac{1}{1} = 1, \frac{-1}{-1} = 1$ and all the answers are in the set.

Mathematics 6-12

65. Which of the following illustrates an inverse property?
(Easy) (Skill 10.1)

A) a + b = a - b
B) a + b = b + a
C) a + 0 = a
D) a + (-a) = 0

Answer: D

Because a + (-a) = 0 is a statement of the Additive Inverse Property of Algebra.

66. Simplify: $\dfrac{10}{1+3i}$
(Average Rigor) (Skill 10.2)

A) $-1.25(1-3i)$
B) $1.25(1+3i)$
C) $1+3i$
D) $1-3i$

Answer: D

Multiplying numerator and denominator by the conjugate gives
$\dfrac{10}{1+3i} \times \dfrac{1-3i}{1-3i} = \dfrac{10(1-3i)}{1-9i^2} = \dfrac{10(1-3i)}{1-9(-1)} = \dfrac{10(1-3i)}{10} = 1-3i$.

67. Which equation corresponds to the logarithmic statement: $\log_x k = m$?
(Rigorous) (Skill 10.3)

A) $x^m = k$
B) $k^m = x$
C) $x^k = m$
D) $m^x = k$

Answer: A

By definition of log form and exponential form, $\log_x k = m$ corresponds to $x^m = k$.

TEACHER CERTIFICATION STUDY GUIDE

68. **Which of the following is always composite if *x* is odd, *y* is even, and both *x* and *y* are greater than or equal to 2?**
    (Average Rigor) (Skill 10.4)

    A) $x+y$
    B) $3x+2y$
    C) $5xy$
    D) $5x+3y$

Answer: C

A composite number is a number which is not prime. The prime number sequence begins 2,3,5,7,11,13,17,…. To determine which of the expressions is <u>always</u> composite, experiment with different values of x and y, such as x=3 and y=2, or x=5 and y=2. It turns out that 5xy will always be an even number, and therefore, composite, if y=2.

69. **Express .0000456 in scientific notation.**
    (Easy) (Skill 10.5)

    A) $4.56x10^{-4}$
    B) $45.6x10^{-6}$
    C) $4.56x10^{-6}$
    D) $4.56x10^{-5}$

Answer: D

In scientific notation, the decimal point belongs to the right of the 4, the first significant digit. To get from 4.56×10^{-5} back to 0.0000456, we would move the decimal point 5 places to the left.

Mathematics 6-12

TEACHER CERTIFICATION STUDY GUIDE

70. A student had 60 days to appeal the results of an exam. If the results were received on March 23, what was the last day that the student could appeal?
    (Average Rigor) (Skill 11.2)

A) May 21
B) May 22
C) May 23
D) May 24

Answer: B

Recall: 30 days in April and 31 in March. 8 days in March + 30 days in April + 22 days in May brings him to a total of 60 days on May 22.

71. Given the series of examples below, what is $5 \not\subset 4$?
    (Average Rigor) (Skill 11.3)

$$4 \not\subset 3 = 13 \qquad 7 \not\subset 2 = 47$$
$$3 \not\subset 1 = 8 \qquad 1 \not\subset 5 = -4$$

A) 20
B) 29
C) 1
D) 21

Answer: D

By observation of the examples given, $a \not\subset b = a^2 - b$. Therefore, $5 \not\subset 4 = 25 - 4 = 21$.

Mathematics 6-12

TEACHER CERTIFICATION STUDY GUIDE

72. **Which of the following is a valid argument?**
    (Average Rigor) (Skill 12.3)

    A) Given: if p then q; q. Therefore: p.
    B) Given: if p then q; ~p. Therefore: ~q.
    C) Given: if p then q; p. Therefore: q.
    D) Given: if p then q; ~q. Therefore: p.

Answer: C

If the premise of a given conditional statement ("if p then q") is true (that is, it is a given), then the conclusion must likewise be true. Answer C, then, is the correct answer. The other options commit various logical fallacies. These arguments may be dealt with by using specific statements in place of the symbols p and q. For instance, let p be "I am in New York City," and let q be "I am in New York." Obviously, if p is true then q must be true as well. Given that "I am in New York City," (that is, given p), then it must be true that "I am in New York" (meaning that q is true, thus demonstrating C to be a valid argument).

73. **When you begin by assuming the conclusion of a theorem is false, then show that through a sequence of logically correct steps you contradict an accepted fact, this is known as**
    (Easy) (Skill 12.4)

    A) inductive reasoning
    B) direct proof
    C) indirect proof
    D) exhaustive proof

Answer: C

By definition this describes the procedure of an indirect proof.

74. **Which of the following best describes the process of induction?**
    (Average Rigor) (Skill 12.5)

    A) making an inference based on a set of universal laws
    B) making an inference based on conjecture
    C) making an inference based on a set of concrete examples
    D) making an inference based on mathematical principles

Answer: C

Induction is the process of making inferences based upon specific, concrete cases or examples. Natural science, for example, uses induction to generalize results from specific observations so as to make broader statements about the characteristics of the universe.

75. **Which statement most accurately describes the concept of a power?**
    (Easy) (Skill 13.2)

    A) A power is repeated addition.
    B) A power is repeated integration.
    C) A power is repeated differentiation.
    D) A power is repeated multiplication.

Answer: D

A power is repeated multiplication. As such, for example, x^5 is equal to $(x \cdot x \cdot x \cdot x \cdot x)$, or the product of five x values.

76. **What would be the least appropriate use for handheld calculators in the classroom?**
    (Average Rigor) (Skill 14.1)

    A) practice for standardized tests
    B) integrating algebra and geometry with applications
    C) justifying statements in geometric proofs
    D) applying the law of sines to find dimensions

Answer: C

There is no need for calculators when justifying statements in a geometric proof.

TEACHER CERTIFICATION STUDY GUIDE

77. **Which of the following is the best example of the value of personal computers in advanced high school mathematics?**
    (Easy) (Skill 14.1)

    A) Students can independently drill and practice test questions.
    B) Students can keep an organized list of theorems and postulates on a word processing program.
    C) Students can graph and calculate complex functions to explore their nature and make conjectures.
    D) Students are better prepared for business because of mathematics computer programs in high school.

Answer: C

Although answers A, B and D may hold some truth, answer C presents the most immediate and profitable value of a personal computer for high school mathematics. Many complex functions are extremely difficult to calculate or represent graphically; personal computers can be a useful tool for faster and more accurate calculation and visualization of such functions.

78. **Identify the correct sequence of subskills required for solving and graphing inequalities involving absolute value in one variable, such as $|x+1| \leq 6$.**
    (Average Rigor) (Skill 14.2)

    A) understanding absolute value, graphing inequalities, solving systems of equations
    B) graphing inequalities on a Cartesian plane, solving systems of equations, simplifying expressions with absolute value
    C) plotting points, graphing equations, graphing inequalities
    D) solving equations with absolute value, solving inequalities, graphing conjunctions and disjunctions

Answer: D

The steps listed in answer D would look like this for the given example:
If $|x+1| \leq 6$, then $-6 \leq x+1 \leq 6$, which means $-7 \leq x \leq 5$. Then the inequality would be graphed on a numberline and would show that the solution set is all real numbers between −7 and 5, including −7 and 5.

79. **A group of students working with trigonometric identities have concluded that $\cos 2x = 2\cos x$. How could you best lead them to discover their error?**
(Average Rigor) (Skill 14.2)

A) Have the students plug in values on their calculators.
B) Direct the student to the appropriate chapter in the text.
C) Derive the correct identity on the board.
D) Provide each student with a table of trig identities.

Answer: C

The personal approach of answer C is the best way to help students discover their error. By demonstrating the correct process of derivation of the appropriate identity on the board, students will be able to learn both the correct answer and the correct method for arriving at the answer.

80.   -3 + 7 = -4   6(-10) = - 60
     -5(-15) = 75   -3+-8 = 11
     8-12 = -4   7- -8 = 15

Which best describes the type of error observed above?
(Easy) (Skill 15.1)

A) The student is incorrectly multiplying integers.
B) The student has incorrectly applied rules for adding integers to subtracting integers.
C) The student has incorrectly applied rules for multiplying integers to adding integers.
D) The student is incorrectly subtracting integers.

Answer: C

The errors are in the following: -3+7=-4 and –3 + -8 = 11, where the student seems to be using the rules for signs when multiplying, instead of the rules for signs when adding.

XAMonline, INC. 21 Orient Ave. Melrose, MA 02176

Toll Free number 800-509-4128

TO ORDER Fax 781-662-9268 OR www.XAMonline.com

FLORIDA TEACHER CERTIFICATION EXAMINATIONS - FTCE - 2008

PO# Store/School:

Bill to Address 1 Ship to address

City, State Zip

Credit card number _____-_____-_____-_____ expiration_____

EMAIL _____

PHONE **FAX**

13# ISBN 2007	TITLE	Qty	Retail	Total
978-1-58197-900-8	Art Sample Test K-12			
978-1-58197-801-8	Biology 6-12			
978-1-58197-099-9	Chemistry 6-12			
978-1-58197-572-7	Earth/Space Science 6-12			
978-1-58197-921-3	Educational Media Specialist PK-12			
978-1-58197-347-1	Elementary Education K-6			
978-1-58197-292-4	English 6-12			
978-1-58197-274-0	Exceptional Student Ed. K-12			
978-1-58197-294-8	FELE Florida Ed. Leadership			
978-1-58197-919-0	French Sample Test 6-12			
978-1-58197-615-1	General Knowledge			
978-1-58197-916-9	Guidance and Counseling PK-12			
978-1-58197-089-0	Humanities K-12			
978-1-58197-640-3	Mathematics 6-12			
978-1-58197-911-4	Middle Grades English 5-9			
978-1-58197-662-5	Middle Grades General Science 5-9			
978-1-58197-286-3	Middle Grades Integrated Curriculum			
978-1-58197-284-9	Middle Grades Math 5-9			
978-1-58197-913-8	Middle Grades Social Science 5-9			
978-1-58197-616-8	Physical Education K-12			
978-1-58197-818-6	Physics 6-12			
978-1-58197-657-1	Prekindergarten/Primary PK-3			
978-1-58197-903-9	Professional Educator			
978-1-58197-659-5	Reading K-12			
978-1-58197-270-2	Social Science 6-12			
978-1-58197-918-3	Spanish K-12			
			SUBTOTAL	
	s/handling $8.25 one title, $11.00 two titles, $15.00 three or more titles			
			TOTAL	